Azure Cloud Adoption Framework Handbook

A comprehensive guide to adopting and governing the cloud for your digital transformation

Sasa Kovacevic

Darren Dempsey

BIRMINGHAM—MUMBAI

Azure Cloud Adoption Framework Handbook

Group Product Manager: Preet Ahuja
Publishing Product Manager: Preet Ahuja
Senior Editor: Arun Nadar
Content Development Editor: Sujata Tripathi
Technical Editor: Rajat Sharma
Copy Editor: Safis Editing
Project Coordinator: Deeksha Thakkar
Proofreader: Safis Editing
Indexer: Tejal Daruwale Soni
Production Designer: Aparna Bhagat
Marketing Coordinator: Rohan Dobhal

First published: July 2023

Production reference: 1050723

Published by Packt Publishing Ltd.
Grosvenor House
11 St Paul's Square
Birmingham
B3 1RB, UK.

ISBN 978-1-80324-452-5

www.packtpub.com

To my friends Julijana (aka Julie), Goran (aka Smile), and Amir (aka an absolute legend), for keeping me sane and all the fun times we have. To my friend Kreso (and his extended family) for all the fun times we have and all the life lessons we learned together. To my mother, Mirjana, for all the sacrifices she made so I can be the man I am today. To all my friends and colleagues at Microsoft, Amazon, Google, and elsewhere for keeping the cloud exciting. To everyone in Ireland, for making me feel as much as home here as I was in Croatia.

– Sasa Kovacevic

To all my friends and colleagues who have been such a great source of support and encouragement over the years; thank you for all the sage advice, positive energy, and boundless enthusiasm – without you, this book would not exist. To my children, Jack, Ryan, and Eva, with their constant inquisitiveness and loose grips; thanks for giving me a reason to learn how to repair many electronic devices. And of course, to my beloved wife, Paula, whose patience and unwavering belief in me has kept me going all these years.

– Darren Dempsey

Note from the authors

Welcome to the cloud adoption journey. With this book, we will explore the myriad of opportunities available to businesses and organizations looking to harness the power of the cloud. For those who are new to cloud computing, it is important to understand the components involved in a successful transition. This book is designed to introduce cloud adoption through Azure, Microsoft's robust cloud platform specifically designed for enterprise-scale projects and operations.

This book will include a detailed exploration of the Microsoft **Cloud Adoption Framework** (**CAF**). The CAF is an integrated set of resources that allow businesses to plan and govern their Azure environment as well as manage costs and ensure scalability. It provides an actionable roadmap that guides customers through every step of their cloud adoption journey—from planning to migration and optimization.

Finally, this book will include numerous anecdotes from real experience of successfully implementing Azure—providing readers with a wealth of practical examples from which they can draw inspiration and guidance during their own cloud adoption journey.

We aim to make your Azure journey smoother than ever before—simplifying the complexities associated with migrating workloads while simultaneously ensuring long-term scalability and that security compliance standards and cost optimization goals are met along the way. We hope you find this book helpful in your journey toward embracing the cloud!

Contributors

About the authors

Sasa Kovacevic is a cloud architect with certifications in all three hyperscale clouds. He loves teaching others and architecting cloud-native solutions as well as innovating how things are done across industries. Sasa's favorite Azure service is SignalR. He can be found on LinkedIn (`https://www.linkedin.com/in/sasakovacevic/`) and Twitter (`https://twitter.com/__S`) and would love to hear from you.

Darren Dempsey is a 25+ year veteran of the software industry, with a career that has seen him architect and develop mission-critical systems for clients in financial services, retail, telecoms, and travel. A self-professed "nerd," Darren's passion lies in solving complex technical challenges, which he has been doing since he was first introduced to computers as a youngster. A keen focus on user experience, system reliability, performance engineering, and keeping things simple drives Darren to continuously innovate and push boundaries.

About the reviewers

Darren O'Neill heads cloud operations at a leading insurance software company, delivering a diverse portfolio of cloud services, including AI, big data, and core components such as security and storage. He leads an expanding team, ensuring reliable support for customers' critical workloads. With a decade of experience in finance and expertise in building secure cloud systems, Darren holds an MSc in cloud computing and an MBA from the National College of Ireland.

Francesco Bianchi is a visual coach and Agile thinker on a mission to make the world of work more fun. His current focus is on building learning experiences in a safe and inclusive environment by leveraging all of the cutting-edge techniques for visual thinking, brain-based learning, and facilitation for highly collaborative meetings to ensure people are and feel at the center.

David Pazdera is a principal cloud and DevOps architect with over 20 years of experience ranging from operations and administration and project delivery to solution design and cloud architecture with a focus on the design and implementation of complex infrastructure environments. He works daily with Azure and IT process automation, cloud governance, and "everything as code." Over the last couple of years, he has been responsible for driving large migration projects to Azure and implementations of Azure landing zones. Before joining Devoteam, he worked for 6.5 years at Microsoft as a senior cloud solution architect. He is also a public speaker with a profile available at Sessionize at the handle `pazdedav`.

Table of Contents

Part 1: The Why

1

Introducing ... the Cloud   1

2

Adopting a Strategy for Success   29

Part 2: The Plan

Part 3: The Execution and Iteration

7

8

9

Closing Thoughts 163

Index 171

Other Books You May Enjoy 180

Preface

With the Cloud Adoption Framework, companies can move beyond just understanding cloud computing and put it into practical use. You'll find a wealth of best practices online – this book helps you make sense of it all! Get ready to learn how to conveniently transition your organization's data onto the cloud while explaining every step of the journey on an executive level as well as a technical one.

Our focus is on the tech and organizational transformation needed for successful cloud adoption. We'll explore ways to build a comprehensive strategy that gets approval, addresses those with reservations, and creates an actionable plan that can drive continuous results even in 2025. Together, we will assess existing practices/processes so they are ready to be replaced once this journey into the cloud begins!

Uncover the tangible implications of cloud adoption and digital transformation to benefit from streamlined, safe operations with fast-paced agility. Dive into real case studies – an enterprise firm, a start-up business, as well as a challenged implementation experience – to craft strategies that align team goals while remaining robustly automated within the cloud environment. Use these insights to achieve peace of mind knowing every interaction is secure without compromising on speed or performance excellence.

Get deep insights into your cloud architecture, diagnose any issues as they arise, and have up-to-date knowledge to deliver measured solutions. Acquire mastery of governance, security, and privacy protocols alongside performance optimization techniques while also knowing when and how Azure services can be deployed correctly to enhance communications within teams or with customers if needed. Organize groups across the organization for better collaboration on uncharted territory – leveraging individual ideas for effective problem-solving strategies!

With this book, you'll have the tools to set your cloud adoption journey on the right path. Carefully selected patterns will deliver a smooth transition experience while avoiding potential pitfalls along the way. By its conclusion, you'll be armed with an intricate knowledge of how best to tackle and succeed in whatever challenges arise throughout forthcoming cloud migration processes!

Who this book is for

Transformative technologies such as cloud computing present a unique challenge for decision-makers. From IT managers and system architects to **chief experience officers** (**CXOs**) and program managers – this book provides actionable strategies for those looking to optimize their organization's cloud adoption journey or get back on course. Chapters provide invaluable insights into how decision-makers can interact with other internal stakeholders in order to achieve success through the power of collaboration.

What this book covers

Chapter 1, Introducing...the Cloud, talks about how you will find this book helpful. It covers the structure guide, learning objectives, and assumptions. You will be introduced to foundational cloud concepts, key Azure services for workloads, app dev, and governance (an introduction to the Well-Architected Framework). It will briefly review **Amazon Web Services** (**AWS**), **Google Cloud Platform** (**GCP**), and application platforms such as Heroku.

Chapter 2, Adopting a Strategy for Success, helps you understand adoption scenarios for enterprises (predominately business IT with large existing investments), start-ups (greenfields with little to migrate), **Software-as-a-Service** (**SaaS**) platform/modern application development (could be a start-up or a unit of a larger enterprise), and common pitfalls (anti-patterns).

Chapter 3, Framing Your Adoption, discusses the Cloud Adoption Framework problem statement (the need), walks through the adoption phases, and looks at alternative approaches (anti-patterns). You will review technology and people readiness, learn how to identify the known unknowns (gaps), and start to understand how disruptive cloud adoption will be.

Chapter 4, Migrating Workloads, discusses building a portfolio/inventory of workloads, reviews Azure migration services, and helps you learn how to assess different options and select the best in line with your strategy.

Chapter 5, Becoming Cloud Native, helps you understand how the cloud enables modern development practices such as microservices and **continuous integration/continuous delivery** (**CI/CD**). It explores opportunities to drive innovation within the organization through greater flexibility, openness, and agility.

Chapter 6, Transforming Your Organization, assesses existing accountability lines across the organization, defines the target state (roles and responsibilities and teams), identifies required skills and learning paths, and redefines **key performance indicators** (**KPIs**).

Chapter 7, Creating a New Operating Model, talks about the beginning of the transformation and shows the path toward the new cloud operating model.

Chapter 8, Architecting Your Cloud, delivers a walk-through of the Well-Architected Framework, covering each pillar in detail, including both the technology and culture dimensions.

Chapter 9, Closing Thoughts, contains conclusions, a wrap-up, and further reading recommendations.

To get the most out of this book

We will strive to align everyone's understanding of the cloud adoption journey, but it is assumed that you will possess a basic knowledge of software architecture, cloud architecture, and organizational change management concepts (albeit with limited experience).

It is assumed that the organization has a general desire to move toward cloud computing and, therefore, already understands concepts such as **Infrastructure-as-a-Service (IaaS)**, **Platform-as-a-Service (PaaS)**, and **Software as a Service (SaaS)**. Additionally, topics such as software-defined networking, Azure services (e.g., compute and storage), SQL versus NoSQL databases, DevOps, and CI/CD should be familiar. We may mention these throughout but will not dive into further detail about them.

Download the color images

We also provide a PDF file that has color images of the screenshots and diagrams used in this book. You can download it here: `https://packt.link/kElDA`.

Conventions used

Following is a text convention used throughout this book.

> **Anecdotes**
> Appear like this.

Get in touch

Feedback from our readers is always welcome.

General feedback: If you have questions about any aspect of this book, email us at `customercare@packtpub.com` and mention the book title in the subject of your message.

Errata: Although we have taken every care to ensure the accuracy of our content, mistakes do happen. If you have found a mistake in this book, we would be grateful if you would report this to us. Please visit `www.packtpub.com/support/errata` and fill in the form.

Piracy: If you come across any illegal copies of our works in any form on the internet, we would be grateful if you would provide us with the location address or website name. Please contact us at `copyright@packt.com` with a link to the material.

If you are interested in becoming an author: If there is a topic that you have expertise in and you are interested in either writing or contributing to a book, please visit `authors.packtpub.com`.

Share Your Thoughts

Once you've read *Azure Cloud Adoption Framework Handbook*, we'd love to hear your thoughts! Scan the QR code below to go straight to the Amazon review page for this book and share your feedback.

https://packt.link/r/1803244526

Your review is important to us and the tech community and will help us make sure we're delivering excellent quality content.

Download a free PDF copy of this book

Thanks for purchasing this book!

Do you like to read on the go but are unable to carry your print books everywhere? Is your eBook purchase not compatible with the device of your choice?

Don't worry, now with every Packt book you get a DRM-free PDF version of that book at no cost.

Read anywhere, any place, on any device. Search, copy, and paste code from your favorite technical books directly into your application.

The perks don't stop there, you can get exclusive access to discounts, newsletters, and great free content in your inbox daily

Follow these simple steps to get the benefits:

1. Scan the QR code or visit the link below

https://packt.link/free-ebook/9781803244525

2. Submit your proof of purchase

3. That's it! We'll send your free PDF and other benefits to your email directly

Part 1: The Why

In today's digital age, adopting cloud technology is no longer a luxury, but we believe it to be a necessity. Organizations that fail to adopt cloud technology risk being left behind by their competition. In this part of the book, we offer a crash course in the cloud and why you should adopt it and offer practical advice and insights for organizations looking to define a successful cloud adoption strategy.

This part of the book comprises the following chapters:

- *Chapter 1, Introducing ... The Cloud*
- *Chapter 2, Adopting a Strategy for Success*

1
Introducing ... the Cloud

Welcome, you absolute legend!

If you consider how many people don't read books, and fewer still read specifically to educate themselves further rather than to entertain themselves, it is a miraculous thing indeed that you are reading this sentence.

The fact that you have chosen this book – hopefully because of a recommendation by a peer either directly or in an online community you frequent – means we were destined to meet, and it means you are already in the top 1% of people in our field who love learning and staying on top of their game.

Thank you for taking the time to read through these pages. Hopefully, we can provide some of the education you seek and add a little entertainment along the way. Now, let us get right to the point as your time is precious – and the cloud won't adopt itself. Let's learn together about everything that needs to happen for an organization (your organization) to successfully adopt the cloud.

Throughout this chapter, you will learn about the authors and why this book exists. We will help you identify gaps in cloud foundational knowledge and guide you to resources that can help you better understand the cloud and ultimately this book.

In this chapter, we're going to cover the following main topics:

- Who are you?
- Who are we?
- What is this book about?
- Are you ready?
- What are the cloud foundations?
- What is the Cloud Adoption Framework?

Who are you?

As any good editor or publisher will tell you, the success of any book hinges on the careful selection of the target audience. We've painstakingly chosen a goal for each and every section, recommendation, and suggestion outlined in this book. But wait, don't worry! That's where you come in. Simply engage your critical thinking skills and soak it all in. Piece of cake, right?

You hopefully take on one of these three personas, professionally:

- *An architect that is going to be leading their organization* through an incredible and exciting change, through a digital transformation (warning: buzzwords).

  Your job isn't just to architect the platform, plan a landing zone, architect services that work together like clockwork, and ensure governance going forward – it is also to ensure everyone in the organization is pulling together toward the same goal: the cloud adoption and transformation of your business and even your industry.

  You are new to this, and you are trying to find your way – and trying to avoid the common pitfalls of cloud adoption.

  You will most likely read from start to finish, skipping only a few sections (for example, a section on migration if you and your organization have nothing to migrate). We'd love to hear from you!

- *An architect that has been through the journey of cloud adoption* and – there isn't an easy way of saying this – things haven't gone to plan. In fact, they have gone terribly wrong, and oh, boy (or girl, or however you want us to address you – hard in a book where we can't hear your replies), do you have stories to tell. Excellent. If you don't put yourself out there and try and give it your best, you won't fail – but you won't succeed either. So, kudos for your bravery.

  You aren't new to this and as some say, *"this isn't your first rodeo."* You have seen things that have worked and things that have failed. Sometimes you made things better and sometimes you were prevented from making meaningful change, but you must dust yourself off and get back in the saddle, *partner*.

  You will most likely pick the most interesting chapter from the list and start reading there and you will be jumping back and forth muttering agreement or disagreement as you read. Either way, we'd love to hear from you!

- *A person in an organization that is going through the transformation process.* You could be the CFO, a QA engineer, or any other role, all this equally applies to your role – some parts more than others, of course. There may be an architect who is leading the transformation and you either have huge respect for their vast technical cloud knowledge or the opposite – you see them struggling and want to help.

  You are interested in learning but need guidance. What are the questions you should be asking? What are the areas you should be focusing on? Who are the people you should be talking to?

You will read this book and go online every once in a while, to learn more about a particular topic. That is to be commended and encouraged.

Either way, your role in this organizational transformation will become a lot clearer after reading this book once, and by the time you go back and read it again to remind yourself of all the great ideas, you should have a chance to implement some of the suggestions. We'd love to hear from you throughout your journey!

If you fit into one of these three personas (heck, even if you do not!), please – we would very much like to hear from you. Tell us how we did, so we can improve. Tell us whether you feel all your questions were answered or you have more questions. Tell us where we fell short and what you would like to learn more about! We'd also love to hear about your own journey, your experiences, your successes, and the troubles you encountered along the way.

Who are we?

Between us… We are a pair of developers or cloud architects, or entrepreneurs, or consultants, or executives, that have had the good fortune to both be in the right place at the right time and have an enormous passion for this field.

They say if you love what you do, you will never work a day in your life (warning: cheesiness).

We've worked across industries; across verticals and horizontals; across industries that are emerging and industries that are heavily regulated; across continents, countries, counties, and cities; working from home or from an office, in open offices, and in our own private offices; across technology stacks and across hyperscale cloud providers; and on-premises, we've been a part of digital transformations before digital transformations were a thing.

We've survived wars, we've survived economic downturns, and we've survived pandemics. We've had experiences we wish we could forget, and we've had experiences that have shaped us and will be treasured for the rest of our lives. Sometimes we've helped others and almost always we've also learned from them. We've dealt with providers large and small, spoken at meetups as small as a dozen people, and at conferences in front of hundreds and thousands of people. We've been a part of many teams and led many teams.

And now, put your hands together for Darren…

I still remember the first time I witnessed someone *programming*. One evening, I watched someone copy BASIC commands from a textbook into a small computer – a Sharp PC (pocket computer), though the exact model escapes me. Looking over his shoulder, I was fascinated by this strange language he wrote that could control a machine and give it a whole life of its own. I thought that, with enough commands, in the correct order, we would magically create AI – I was 8 years old, and the scale of such things was beyond me. Still, this led to owning a computer, learning to code, and spending the majority of my time since then in front of a computer screen. But I regret nothing!

I stumbled across AWS sometime around 2008. Initially, I was dismissive of S3 – to me, it was just web hosting, but with better marketing. Then came EC2, but VMs were nothing new. Local web hosts had been offering these for some time, but Amazon could do it at a fraction of the cost. Interesting… But with SQS, things got really interesting. A message queue, the backbone of any enterprise application could be created with a credit card and a few clicks. But would any serious organization trust a bookshop with even a tiny piece of their IT infrastructure?

Well, we all know what happened next. With more and more services from AWS, more players entered the space to copy and compete with Amazon. But few dared to imagine how the cloud would radically alter the IT landscape and I would wager fewer still could have predicted the effect it would have on the culture of IT departments and technology companies globally.

I feel incredibly lucky to not only have witnessed such rapid transformation in technology over the last three decades, but to actually deliver solutions that delight users – all built on the cloud created by the most innovative technology companies that have ever existed.

And once more, put your hands together for Sasa (read as Sasha)…

I taught myself how to code in a basement shelter in third/fourth grade during the Croatian War of Independence, with no electricity, reading four books on GW-Basic. I coded my first game on paper before I ever saw a computer – I had to simulate random numbers by annoying people in the shelter to pick a number (by the way, one of the worse ways of generating randomness).

I started working as a developer and I still am a developer, even though nowadays people sometimes call me an architect.

I've worked in public and private sectors, telecoms, financial institutions, health care providers and insurers, defense, law enforcement, secret services, and governments. I've used Java and C#, Python and bash, SQL and NoSQL, and Windows and Linux (and Solaris). I was there when AWS started with my favorite service, to which my best friend introduced me – **Simple Queue Service (SQS)**, and I was there when Azure launched my now all-time favorite service – SignalR (if you don't know anything about it, drop this book now and go learn about it – everything in this book can wait; SignalR is just so useful as a service). I am also a massive fan of Google's cloud efforts because I feel we cannot have just two providers dominate the market as that will lead to trouble. (We are already beginning to see some of this troublesome behavior.)

I've been called on to approve deals worth over $50 million and I've helped the digital transformation efforts of global, strategic top-500 customers of Microsoft. I have talked to C-level executives and developers and quality assurance engineers and project and product managers – and I have successfully (sometimes through many iterations) convinced them to follow a course of action that led them to get the full value out of their cloud investments.

> And now, to answer the inevitable questions that always come up – AWS versus Azure? Azure wins for me for two reasons: the Azure portal is amazing when compared to the AWS Management Console, and Microsoft account teams (both commercial and technical) will do anything to help you (as will AWS') – but the sheer commercial and enterprise power of Microsoft is unrivaled if you are looking to partner commercially. AWS has a better support organization, in my humble opinion. And Google – they should be doing a lot better than they are (maybe they could use some help?) in the cloud wars, but you can definitely run your services in any of these three clouds – if you know what you are doing. Do you? Do you know?
>
> The easiest way to contact me is at `linkedin.com/in/sasakovacevic/`.

Enough of the fancy words, enough of the introductions; it's time to tell you exactly how this book will help you – today.

What is this book about?

Very specifically, this book will help you with cloud adoption by describing the following exactly:

- What you need to know to be able to strategize, plan, govern, manage, and innovate your cloud adoption and your applications and services in the cloud (with examples and focus on Azure while still being applicable to any cloud or any multi-cloud environment).

- Who will be your friends in a constant struggle to stumble forward with agility and confidence, how to manage relationships and activities with your friends, and how to be the evangelist that sees every interaction and every touchpoint with anyone in the organization and outside of it as an opportunity to include them and bring them along on the cloud adoption journey. Repetition is key! Repetition is key!

- How to get things done in an iterative and agile way and with one eye (or ear, or finger) on the business needs and the other on the technical requirements.

- When is the right time to approach each topic, when is the time for compromise, and when is the time for decisive action to achieve the business goals.

- The anti-patterns – things that may make sense initially but have been tried and proven not to work.

This book (if you are still following what we were discussing) will equip you with the tools to use with this practical guide to define and execute your cloud adoption strategy. But every business organization is different, at different stages of maturity and with different ideas about what success looks like.

You will walk away from this book with knowledge, specific insight, and a practical plan (and a mindset as well) that will help you and everyone in your organization define and execute a cloud adoption strategy.

We will explore a wealth of past experiences that have enabled us to deliver smooth execution of cloud migrations. We also want to highlight areas that are ripe for innovation.

Industry-specific considerations such as compliance and data security will be at the forefront. We won't focus as much on a specific technology or go in-depth on how to use it but try and set broad standards and focus on technology as it enables organizational transformation.

We will also investigate the organization's transformation and how to achieve it, who you absolutely must bring along for the ride, who will go willingly, and who you will have to drag kicking and screaming into the cloud.

You will learn how to create a compelling strategy that gets buy-in across the organization, and which approaches work to win over and influence those with the most to lose (and how to have them look at the wins that are available to them).

No plan survives the first contact with an enemy on the battlefield, but forming a realistic plan is a must for you to be able to deliver and govern cloud adoption and the cloud itself over the long term (or at least for the next 2-3 years before you change organizations).

You will also learn how to recognize the right time to, and how to, decommission existing practices, processes, and technology, and replace them with those appropriate for the cloud in 2025, 2030, and beyond (of course, being mindful that long-term plans are just a beacon in the fog of uncertainty).

The plans you will make need to be general, broad, and adaptable in order to be able to survive contact with the enemy (that is, market forces). General plans that cover a broad range of circumstances are better than specific, narrow plans. You will also need to understand which services are not going to last (in our opinion), which services are vaporware, and which technology trends are important for cloud adoption and your industry.

Finally, we will also give you some tips and you will learn all about navigating cloud adoption in heavily regulated industries such as finance, insurance, defense, and so on.

By the way, a convention to be on the lookout for...

Throughout this book, you will come across sections such as this next one, where one of us (the authors) will interject with an opinion or an anecdote, asides, and tangents that will briefly, or at length, describe something you might want to investigate after you have read, understood, and started implementing the wonderful things you've learned here in this book – so be on the lookout for them.

Here is an example of an aside – one of the many questions we get a lot.

Multi-cloud? Yes or no. Or, when?

Absolutely never, *except* if you are an organization with thousands of developers and have products that are not all interconnected; if you are an umbrella organization and have acquired companies that are already using different clouds and have customers in production; or if you are in a regulated market or a government entity with mandates for multi-cloud.

Regardless, if you can choose, do not choose multi-cloud. You must double or triple governance and you limit the growth and cross-pollination of services and developers. Also, it's a huge pain adopting one cloud. Why adopt more than one if you really don't have to?

If you must go with multi-cloud, then pick one primary cloud, do well with it (that is, adopt the hell out of it), and only then introduce the second one. Stay away from three. There be dragons.

Simple really.

But, but, but… what about vendor lock-in? That is not a thing, in so far that everything you do locks you in, so stop worrying about a future issue that may never come up, stop focusing on the lowest-common-denominator technology, and embrace – adopt – the cloud. If you must do multi-cloud, adopt one well and then introduce another.

And, again – in the next edition of this book, you could see your anecdotes here as well, so contact us if you have something to share. We'd love to learn from your experiences and share them with future readers.

I once had a client…

I just want to make sure that if you have been through this journey and had issues along the way, you understand that these things happen to the best of us.

Cloud adoption is hard on both the technology and business levels. Sustainable, governable, and painless cloud adoption is a rare exception – one that we want to help you replicate here.

So, I once had this client (one of many), a huge global financial institution that had attempted to adopt the cloud as best as they knew how and had unfortunately failed spectacularly.

They failed so spectacularly that the regulator fined them and made the governance processes so stringent and hard that they had to completely scrap their effort. And they had applications, services, and customers – live in production. But the mess was such that only a hard reset and only a change at a VP level could get them out of this crisis. Whole departments were disbanded and the organization had to undergo a re-org, and then another one just for good measure, to be able to start again.

This time, in a much smarter (and a reasonably cautious) way, with buy-in from every level of the organization. And they are just now, after a year and a half, coming back with those applications, services, and customers – to production.

Some failures you accept and shy away from, and some you embrace and you do better – maybe with some outside help.

This is one of those double, good news/bad news types of situations.

The bad news: they wasted a lot of time, their competitors plowed ahead, and they suffered in the process. The good news: they understood the benefits of the cloud and were still very keen to try again, and they are now doing a lot better having understood that cloud adoption is easy to do poorly and hard to do well – but well worth the effort.

If only there was a book they could have referred to in their time of need, or if only they had good people that read such a book and understood the complexities of a complete digital transformation. If only…

We've met you now and, hopefully, you now understand that this book was tailor-made for you.

Are you ready?

Change is hard. Changing an entire organization is even harder. It is made harder still when coupled with such a huge technological paradigm shift as cloud computing, which not only requires knowledge of software, networking, and cloud services but also the knowledge of high-level concepts that cloud computing brings to the forefront.

Let us share a hard *truth* with you: almost every organization in the world is now adopting the cloud or planning to adopt the cloud – and they are all trying to do it as a matter of course, as a thing you do and complete and get done, as a thing you do as you've done before. Doing that is not impossible, but the results from such a strategy are lackluster at best. Take heed and bear witness to the truths that lie herein (to quote the tales of the *Horadrim* from the computer game Diablo) – nothing short of revolutionary organizational change and acceptance that we are not in Kansas anymore (to quote from the Wizard of Oz) is going to suffice.

You can either *accept* the need for this *or* you can try and *fight* it, do what you've always done, and stick with your traditions of IT change management. Hopefully, it is slowly dawning on you how seriously you and your organization must take this process. For nothing is at stake here, other than the very future of your organization.

The cloud architect is the one person in an organization that needs to understand all of this. Must. Understand. All.

They must be able to convey the importance of each topic to others in the organization and will need to work with all levels in the organization to bring about the change.

 In cloud computing, the top priority is to achieve the business goals of the organization. All other matters take a back seat. Focus on what really matters. And in this book, we assume that the business goals are broadly aligned to do the following:

1.   Deliver digital services to customers (internal and external).

2.   Be quick but diligent about it (agility and compliance).

3. Pay for agility and acceleration but don't break the bank (focus on agility first, but then circle back to cost optimization regularly).

4. Enforce the brand and the reputation of the organization (security and sustainability).

5. And lastly, have peace of mind and the time to learn new things (less firefighting, more innovation).

Your organization has decided to go all-in on the cloud. So, the buy-in, in principle at least, is there. Now, how does one transform the organization to be able to quickly and efficiently deliver on that promise in its day-to-day operation.

This book will address this challenge by showcasing the actual path to take from day zero (the decision) to strategy formation, planning, and execution, all the way down to day-to-day operation and long-term management. Anything short of total organization and technology transformation will miss all opportunities the cloud provides.

Agility: Business needs must be addressed yesterday, not in six months or two years. How does the adoption of the cloud help address this? What in the organization is preventing innovation, faster time-to-market, cost efficiency, and global scale? Not just one agile team, but repeated over and over again, at all levels and across all teams, in an orderly, organized, governed, and compliant way – without adding more bureaucracy, but rather by empowering all levels of the organization to be agile by default.

True agile adoption requires one to steer the organization not with slight and sporadic nudges but through focused radical course correction. Imagine a massive oil tanker attempting a 180-degree turn: the process is slow and appears to only make small incremental changes, but it is predictable and when it starts there is no stopping it.

This book will address this challenge by providing callouts, funny anecdotes, adoption stories from enterprise and start-up perspectives, information from running a SaaS platform in a regulated industry, ideal and cynical views, examples of what did work and what didn't, patterns and anti-patterns, and so on.

Agility

I cannot stress this enough, and I've had this conversation with many CFOs, product and project managers, and even developers. To quote Donald Knuth, *"premature optimization is the root of all evil."* This applies to culture as much as it does to code. Trying to optimize your practices while trying to achieve agility is lunacy – you don't know yet what is important, or where the bottlenecks will occur.

Focusing on agility means focusing on the easiest path to production. If that means procuring more expensive services either on a higher tier or at a larger scale, do it. You do not want to waste time arguing about the sizes of VMs, App Service plans, or do we use Service Bus or Event Hub. Repeat after me: *It doesn't matter.*

Remember that you are trying to develop, deploy, and deliver a digital service to customers. How do you know if it is a success or a failure? You get it out there, into the hands of your customers (again – external or internal), and you gather telemetry and feedback.

Is the service performing its business function and bringing value? Now push on and deliver more. Once you are getting diminished returns on the new features, go back and focus on optimizing the service. You may have paid for a few months of more expensive tiers and services and some of them may not have been optimal, but they were being used and the business is better for it.

The only exception to this is security: you do not compromise on security – ever.

And if the service wasn't successful, evolve it from telemetry and feedback – or, kill it with fire. Be ruthless. You must. Or the market will be ruthless.

One practical example of this was the delivery of a COVID-19 vaccination registration and scheduling form. We did not compromise on security, but we picked the easiest (fastest, most agile) services to be able to get the form into the hands of the customers as soon as possible.

We ensured elasticity and when 5 million people accessed the form on day 1, it just worked. Then it took us two weeks to move that to more cost-efficient services and evolve the service further (for example, to be able to amend the scheduled slot). We paid for a few weeks more than was necessary, but the service was live and in the hands of those that needed it.

We could have waited a month and done the optimization upfront and then rolled it out, but you can get a lot of people vaccinated in a month.

Another example might be a service for the world's largest sneakers manufacturer. The decision was between optimizing for agility and deploying a service that would be ready for Black Friday and the Christmas season or optimizing for cost and deploying the service in time for Valentine's Day.

After stating it in those terms, which path do you think they chose? Which path would you choose?

Practicality: An architect needs to be in control of everything from innovation to workload deployment, scalability, agility, governance, and so on. It is literally impossible for one person to mind all these things, so to scale, an architect needs to influence the rest of the organization to get buy-in initially and continually, and to help create an organization that is then by default ready to address these challenges without the need to micromanage, argue, or struggle to deliver on all levels of the organization.

This book will address this challenge by providing the patterns and mechanisms (and sometimes just pure practical advice) on how to achieve this goal. This is a continuous process that is overwhelming initially and like any new process, it is initially painful as it involves all levels of the organization, but with the proper strategy and planning it can be done at scale.

Understand cloud adoption and digital transformation generally, and what it means in practicality in the day-to-day running of a cloud platform. Learn from actual examples – an enterprise company, a start-up/greenfield site, or a less than successful cloud adoption.

Be able to plan the cloud adoption journey and help all levels of the organization to do so as well. And then execute on that.

Innovate with the business goals in mind, then execute with cloud workloads that are automated and deployed in a predictable and safe manner, in a fast and agile way, without worrying every time someone interacts with the cloud that something will go wrong.

Have an overview of the entirety of your cloud workloads, what they do, why they are there, how they interact with each other, and how to deal with any issue relating to them being there – from communicating internally on required improvements to communicating internally to stakeholders and externally to customers when things inevitably go wrong.

Understand something about the concepts of governance, security, privacy, reliability, operational excellence, cost optimization, performance efficiency, and a whole bunch of Azure services and how/when to use them. Organize teams across the organization and join any of those teams to showcase your ideas or help them understand a difficult cloud concept.

Assumptions

Top of the list of assumptions is your organization has or is planning to have a cloud-first strategy and you have a significant role to play in it – hence we've written this book for you. So, we assume you have some understanding or maybe some experience of concepts such as these:

- Infrastructure as a Service
- The major cloud providers: Microsoft, Amazon, and Google

We will attempt to bring everyone to the same level of knowledge, but in general, we assume that architects have general knowledge of (but may lack extensive experience with some or all of) software architecture, cloud architecture, and organizational change management concepts.

Cloud foundations

As we've said before, this part of the book will be a level set for everyone to be on the same page with the basic concepts, so if you are 100% sure you understand the following, you can skip this part:

- Basic cloud concepts
- Security and privacy implications
- Cloud services
- Cloud workload types
- Pricing and support options

If you feel less confident, maybe just skim this section. You can always come back and read it if required. Also, when you gift copies of this book to the people in your organization, they can quickly catch up with acronyms, concepts, and ideas here. If you are one of those people that got gifted this book and need to understand the cloud concepts, welcome! Someone in your organization loves you enough that they would like you to educate yourself further, to be an active participant in your organization's digital transformation journey.

> **Smash that like button!**
>
> I feel like I should now shout at you to comment, like, and subscribe as it really helps the channel out. But this is not YouTube, so it might be a bit more difficult for you to do. So, get in touch in other ways!

As we continue forward, we will focus on Azure services, however, these concepts and the concepts in this book more broadly apply to any hyperscale cloud provider, so if you are primarily working day to day with AWS or GCP, you should be perfectly fine translating these to their respective services.

Cloud concepts

So why adopt the cloud and why should we care for it?

Again, you really must think of it in terms of agility. The cloud is a way for us (all of us) to deliver value into the hands of our customers faster. It is also a way for us to deliver value that we just couldn't before from our own data centers. Be it due to physical constraints or economical constraints (hundreds of thousands of compute units at our fingertips), it was not easy or quick to run an experiment or prototype an application, let alone test it with a small subset of our customers and then scale it to the entire global market.

The cloud also brings services that we would normally have had to introduce ourselves into our architectures and plan for their development, deployment, testing, scaling, supporting, updating, monitoring, and so on.

Azure Service Bus

A service such as Azure Service Bus now gives us the flexibility to handle publish-subscribe events with one deployment of a template and all our services can avail of it, without us having to develop, deploy, and test it for functionality.

It can be highly available and scale automatically, it has 24/7 support, and it updates itself both in terms of bringing additional functionality and security (it can even be made highly available across Azure availability zones by just picking the premium tier and configuring it) and by delivering new features. That is an awful lot of work we don't have to do. Focusing on features and services that your organization cares about, you shouldn't be building a service like Azure Service Bus.

Azure Service Bus is a comprehensive offering with many options and possible configurations. There is a wealth of information available on the Azure website on things such as pricing tiers, messaging design pattern scalability, observability, and so on. For example, try out these links:

- Azure Service Bus: `https://docs.microsoft.com/en-us/azure/service-bus-messaging/service-bus-messaging-overview`

- Publish-Subscribe pattern: `https://docs.microsoft.com/en-us/azure/architecture/patterns/publisher-subscriber`

- Azure support: `https://azure.microsoft.com/en-us/support/plans/`

Azure Monitor

As another example, a service such as Azure Monitor brings with it a whole wealth of integrations (automatic, semi-automatic, and manual) that allow you to monitor your entire Azure estate from a single pane of glass (to use another buzz phrase there). This means that for all Azure services and for a whole bunch of applications and services you are building, you get out-of-the-box monitoring and metrics without you having to do anything other than configuring the endpoints and start ingesting your telemetry.

The power of Azure Monitor (the App Insights part of it especially) doesn't end there as you can extend Azure Monitoring default events with your own custom events, which usually Azure cannot reason about on its own – for example, every time a person completes a level in your game, evaluate all the inputs, game paths, times, and scores, check them for cheating and submit an event into App Insights on the result of the evaluation. Later, you can investigate these events either automatically or manually and further improve your anti-cheat methods.

Different definitions of the cloud

Getting back to the concept of the cloud, you now understand why the cloud is so powerful. But now let's switch to what the cloud is. Ask 10 people to define the cloud and you will get at least 13 answers. Ask these 10 people tomorrow and you will get 13 completely different answers. And for sure at least 50% of all those are correct. They might even all be. The cloud means different things to different people. What does it mean for you?

CFOs might focus on cost-saving provided by PaaS services over IaaS and traditional virtualization in their own data centers – bringing costs down means an opportunity to reinvest in more research.

CTOs might focus on the standard catalog of services to be used in a compliant and repeatable way – henceforth bringing an easy onboarding of future services the organization creates.

The head of engineering might focus on reusable components, technologies, and services – thus unlocking career progression opportunities for team members to move between different teams with ease.

A developer might focus on writing just the code they need – rather than also needing to worry about what type of infrastructure will be needed to run the code. They also might focus on how easy it is now to debug in production when compared to on-premises deployments in customers' environments the developer had no direct insight into.

And all of these are true, so how can the cloud bring that about?

What is the cloud?

Picture a seemingly endless web of physical servers spread across the world. These servers, each with their own special tasks, come together to create the all-encompassing wonder we call the cloud. The cloud is compute, storage, memory, and common IT building blocks at your fingertips without the (traditional) headaches. It is also, for most purposes, "infinitely" scalable in those dimensions. (Alright, technically not infinite, but we rarely worry about having the resources available to scale typical business applications.)

The cloud also delivers global scale, massive bandwidth, and minimal latency through data centers located closer to your customers than you can ever be. Azure has more than 60 regions with more than 60 data centers and tens of thousands of edge locations (with partners such as Akamai and others).

Could you build a service and offer it globally and cheaply before the cloud? Sure, you could. Would it be as cost-optimized? Hell no! Can you do so today in the cloud literally in hours? Yes, absolutely. You can and you should.

The cloud is a vast network of virtualized services ready for you to pick and choose (cherry-pick) which ones you need and is capable of bursting and scaling as you require for your services. The cloud is glued together by an unimaginable length of wires, numbers of chips, and ultimately is a testament to human ingenuity and a vision of evermore powerful computers in the hands of every single person and organization, enabling them and you to achieve more – every day. I hope Microsoft forgives the paraphrasing of their corporate mission here: *empower every person and every organization on the planet to achieve more*:

Figure 1.1 – Cloud meme

The cloud is, as that meme tells us, someone else's computer. In fact, it is hundreds of thousands of someone else's computers, and even millions of them. And it's perfect that it is that way. We can use them and we don't have to maintain them.

> **I don't own a car**
>
> I rent. Either long-term (weeks and months) or short-term (hours or days). I get the benefit of a car – transportation. I don't get the hassle of repairing and maintaining the car, taxing and insuring it, and a piece of mind worrying about what happens if I scratch it or crash it. I get any car that I want, small and cheap, large and useful, or fancy and expensive.
>
> Is renting a car for everyone? Maybe not – self-driving cars may eventually bring about a mindset change for us all. Is this marginally more expensive on a per-use basis? Yes. Are the benefits worth it? Absolutely. Do I even like driving and am I even a good driver? No, to both. Am I terrible at parking? Yes. Did I get to drive cars I would never be able to afford (at least before everyone and their friend buys this book)? Yes.
>
> Like the cloud, where you rent compute, memory, storage, and bandwidth, I rent cars.
>
> And both renting computers in the cloud and renting cars are a future certainty. An inevitability that is coming soon for all of us.

Note that if you come across any unfamiliar terms, you can probably find a brief description within the Cloud Computing Dictionary:

`https://azure.microsoft.com/en-us/overview/cloud-computing-dictionary/`

Now that we are aligned on the cloud itself, let's focus on to what it means to architect for the cloud.

Architecting the cloud

The common design concerns for cloud architecture are these five pillars:

- Operational excellence
- Reliability
- Performance efficiency
- Cost optimization
- Security

Operational excellence is defined by Microsoft as operation processes that keep a system running in production. That is a very narrow definition.

What you should consider under the topic of operational excellence is why the processes are set up such as they are in your organization and what needs to change to achieve agility. You should also look to balance your team's freedom (the desire to do as they like and define their own processes) versus following standard processes as defined.

There are two key high-level concepts to understand here – **Application Performance Management (APM)** and **Infrastructure as Code (IaC)**/automation.

APM tools must enable visibility of all aspects of application performance, from server health to user experience, giving teams deep insights into how their applications are operating in real time. APM, over time, should provide teams with data points of how changes are impacting their applications, positive or negative, allowing them to take proactive measures, for example, to maintain an optimal level of efficiency or pivot on the functional direction of their application – this type of agility is core to operational excellence.

IaC and automation go hand in hand. They essentially mean that nothing you do should be unique, one of a kind, or manual. Every change and every action needs to go through continuous integration and continuous deployment pipelines. It needs to go through as a unit, as a completed task that is traceable from the idea to the line of code to the telemetry event. This needs to be repeatable and must produce the same identical result every time (this is also referred to as *idempotency*).

What this also gives you is – say it again – agility. You must be able to roll back any change that is causing disruption, performance dips, or instability.

Is that easy? No.

Is there a lot to do to prepare for that? Yes.

Can it be done iteratively, so we get better at operational excellence over time? Yes.

Is it worth the peace of mind achieved once it is in place? Yes.

The end goal is for you and for everyone in your organization to be able to step away from work at any time for vacations, for fun and adventure, or just to sleep and have no worries that anything will happen that won't be automatically resolved (by at least rolling back to what worked before). If your organization can deploy new code and new services on a Friday afternoon and no one cares or worries, you are there – you are living the dream of operational excellence. If you are one of these individuals, we'd love to hear from you.

Have I seen any organization achieve all of this? No. Never. Some, though, are so very close.

And that is what it's all about – doing better today than you did yesterday. And every good deployment, and equally every bad deployment, is an opportunity to learn. No one needs to accept the blame and no one needs to get fired – the solution is always that the process improves.

Yes, someone may still get fired and even prosecuted for deliberate malicious activity, but the solution is and must always be the process improves, we improve, and we do better going forward.

Reliability is defined by Microsoft as the ability of the system to recover from failures and continue to function. The key thing is to continue to function.

DDoS

I've had customers work with me and try and work out what they do with their services if a DDoS attack is initiated against them. Inevitably, someone will mention we should probably just turn all the services off to save costs in the event of DDoS as throwing infrastructure resources at the problem is sometimes necessary, so just shut down the services and wait until the attacker goes away.

To which my reply is always, let us consider the reason behind a DDoS attack and what the goal is. Pause here and think. What is the goal?

OK, so if the goal is to make your services inaccessible to others, what good does shutting them down do, except doing exactly what they wanted to achieve? For example, a DDoS attack against an Xbox service is designed to make gamers unable to, well, game. If you then turn off the service as a response, what have you achieved?

The key thing about reliability is for the services to continue to function.

DDoS mitigation could very well be a book in its own right so we won't go into that here, but just to give you a head start: Azure has a service that mitigates DDoS attacks, one tier being free and the other costing you money. Turning that on is a really (really, really) good idea for public-facing endpoints. Also, Microsoft will have teams ready to assist at a moment's notice if the attack does happen and the automatic mechanisms don't prevent it. And you will have a priority service if that is the case.

So, how do you achieve reliability in general terms? Well, primarily with redundancy. You ensure no single point of failure exists within your architecture.

Business requirements

Before you invest time in high availability and resiliency from a redundancy perspective, ensure that is the actual business requirement. I've seen so many teams struggle to achieve unreasonably high availability, only to answer my question "*What is the traffic to the service?*" with "*Nine queries a week on average.*" Or, my question "*What exactly does the service do?*" with "*PDF generator*". Unless your business is PDF generation, people can usually come back for their PDF or wait until it is processed and generated in a background thread and emailed to them.

I am already looking forward to all the feedback like "*Well, actually, our PDF service is mission-critical.*" All I am saying is think before you invest effort in reliability. Ask the business how critical the service is.

And another aside here: if all services are critical, then no service is critical. This has a slight possibility of being incorrect, but I've never seen it.

Another way to improve resiliency is for the services to fall back to less-intensive responses. For example, if the service returns the most sold items today and it needs to do massive parallel queries and correlate the values, it can fall back to a cached list from yesterday, which is just one fast query.

Resiliency is another topic we could spend a lot of time on, but for now, just remember these concepts: single point of failure, graceful degradation, and one last thing – if there are issues with one service in your architecture, expect issues to cascade up and/or down the stack, and even after you have mitigated the issues, expect further issues in the next week or two, so be prepared and staffed. A rule of thumb– here for you free of charge (almost) – will save you a lot of headache.

The reason behind this is that in architectures we see today, interconnectedness is baked in (unfortunately) more than it should be as it is often not easy to visualize all the dependencies, so maybe work on visualizing those as well – before issues happen.

Why is it that in the cloud, which is so powerful and useful, these issues are more pronounced? Well, there are now more people and machines connected to the internet and there are more and more services being used by more and more people and machines, so this wasn't such an issue in the 1990s, but it is today. The underpinning concept behind cloud computing is using commodity hardware, and at such a scale that small percentages matter. For example, 1% failure per year on 2 disks means disks will be fine almost all the time. But 1% failure at a scale of 60 million disks means that 600,000 will fail this year. That is an issue. And while disks fail at more than 1% per year, other components must also be considered, such as chips, and so on. Also, the cloud is, for our purposes, public (as opposed to the private cloud), meaning the cloud is a shared service. Though logically isolated, you may find yourself with noisy neighbors that may impact your services. You will get hackers from the bedroom variety to the state-sponsored type that sometimes do, but most often don't, target specific organizations, but rather spray and pray they get you – and you too can pray that you don't get caught in the crossfire.

Now that you are in the cloud, you also need to consider that updates to the underlying technology don't always go well, and Microsoft, Amazon, and Google will destroy and disrupt services in one or all regions, with regularity. No slight meant here against their SRE teams, that is just again playing with large numbers and small percentages. If they do 1 update a year, then 1% failure is negligible, but if they do thousands a day, then 1% starts growing rapidly. However, that is the whole idea behind the cloud – everything can and will fail, and you will learn to love it and understand it because that very fact brings about new ways to simplify and plan for high availability in a different way than if you were running your own data center. Not to mention you and your organization are not above failure as well.

> **Risks**
>
> What is the risk of losing a data center? I have seen risk logs with an entry for a scenario were a meteor crashes into the data center. But that is such a remote chance that your cloud provider destroying a data center is much more likely.

Now that you know that failures are not only expected but inevitable, you can design and architect your services around that – if the business requirements are there that demand it. Remember, people doing manual work make so many more mistakes compared to automated machine processes – hence automation is again your friend. Invest in your friend.

Performance efficiency is defined by Microsoft as the ability of the system to adapt to changes in load. And this again brings us back to… agility. How hard do you have to work for your service to go from supporting one customer to a billion customers?

Can you design and configure a service that does this automatically? Azure Active Directory, Azure Traffic Manager, and Azure Functions are examples of such services with auto-scaling.

Prefer PaaS and SaaS services over IaaS, prefer managed services over unmanaged, and prefer horizontal scaling (out) rather than vertical (up). This also applies to scaling up and down.

You should consider offloading processing to the background. If you can collect data and then process it later, the performance will improve. If you can process data not in real time but as and when you have the baseline capacity, the performance will improve.

You should consider caching everything you can – static resources. Then consider caching more – dynamic resources that aren't real-time sensitive as well. Then consider caching more – results from the database, lookup tables, and so on. When should you stop caching? When everything is cached, and you can cache no more, the performance will improve. A great caching service is Azure Redis, but it is by no means the only one. Another amazing one to consider is the CDN service.

Have you considered your write and read paths and are they stressing the environment? Try data partitioning, data duplication, data aggregation, data denormalization, or data normalization. All of these can help improve performance.

Cost optimization is defined by Microsoft as managing costs to maximize the value delivered. The concepts we've discussed previously go hand in hand with cost optimization.

Are you using the most expensive service to store data? Azure SQL is great when you need queries, and you need to do them often. But having a 1-TB database for the past 6 months of records that keeps growing while all your users only search today's events is a waste.

Moving data around is what you should get used to. Use the right storage and the right compute resources at the right time. Moving data to another region may be costly but moving it within the region may be completely free. And using the most appropriate storage can save you millions. And to facilitate this, a lot of Azure services provide data management and offloading capabilities.

Cosmos DB has time-to-live functionality, so if you know an item won't be needed after a time, you can expunge it automatically, while you can still simultaneously store it in a file. Azure Blob Storage has Hot, Cold, and Archive tiers and it can move the underlying storage automatically as well. If the file is no longer needed to be highly available, move it to lower-tier storage – you will pay a lot less.

And remember, there is an egress cost! When you are about to move data, always ask *What about egress costs?*

Security, as defined by Microsoft, follows the zero trust model in protecting your applications and data from threats – including from components within your controlled network. There are so many ways to protect your workload.

We have Azure DDoS Protection, which protects against denial of service attacks; Azure Front Door geo-filtering, which limits traffic that you will accept to specific regions or countries; Azure Web Application Firewall, which controls access by inspecting traffic on a per request basis; IP whitelisting, which limits exposure to only the accepted IPs; VNET integration of backend services, which restricts access from the public internet; Azure Sentinel, which is cloud-native **security information and event management (SIEM)**, and so on.

A lot of these don't require you to manage them day to day – you set and forget them. For example, with VNET integration, once you've enabled it and written some automated tests to ensure it works every time, you are done.

These pillars come from the Azure Well-Architected Framework, which goes into a lot more detail that won't make it into this book (although we will dig a bit deeper later):

- Microsoft Azure Well-Architected Framework:

  `https://docs.microsoft.com/en-us/azure/architecture/framework/`

  You can also find self-assessments there that will help determine your current maturity and future work ahead of you.

- Microsoft Azure Well-Architected Review:

  `https://docs.microsoft.com/en-us/assessments/?mode=pre-assessment&id=azure-architecture-review`

AWS and GCP offer similar guidance as well. These are specific to each hyperscale cloud provider and to each service and concept as it pertains to them, so while the general concepts are similar, the actual guidance may differ based on service definitions and implementations.

Cloud security and data privacy

Security is a shared responsibility between your entire organization and your cloud provider. Especially, as we are playing here on different levels, from the physical security of the data centers to the security of your passwords and other cryptographic secrets you need in your services' operation.

You need to protect your – as well as your customers' – data, services, applications, and the underlying infrastructure.

Services such as Microsoft Defender for Cloud are your friend and will give you plenty to concern yourself with – everything from ports open to the public to automatic security insights such as traffic anomalies, for example, machine A has started communicating with machine E and has never previously done so.

You will also need to understand the patterns around the use of Azure Key Vault and how to successfully use Key Vault in your IaC scripts and in your applications and services.

Then there are services that protect the public perimeter, such as Azure DDoS Protection, Azure Front Door, Azure Application Firewall, and so on. And each service has security recommendations and best practices and guidance on how best to protect it from internal and external threats.

Sometimes though, you will just need to guarantee that data hasn't been tampered with, so we slowly start moving from security to compliance. **Azure confidential ledger** (**ACL**) is one such service that ensures that your data is stored and hosted in a trusted execution environment. The scope around these is fascinating and the science and patterns are really showcasing what is possible today with technology – not just possible but guaranteed.

In Microsoft, there are teams whose job is to ensure the compliance of services and the platform with legal and regulatory standards around the world. You name it, they have it. AWS and GCP are close behind as well.

Compliance Portal:

`https://docs.microsoft.com/en-us/azure/compliance/`

Again, a reminder that implementing recommendations from any or all of these does not mean you are compliant as well or that you are secure. Shared responsibility means you still must do your due diligence and work to satisfy the requirements of compliance frameworks. Theory and practice both must be satisfied.

Cloud services

As mentioned, we've focused on Azure in this book as a primary hyperscale cloud provider, but here are three great pages (one from GCP and two from Azure) that give an overview and compare services and offerings so you can easily understand similar services across these providers:

- AWS, Azure, GCP service comparison: `https://cloud.google.com/free/docs/aws-azure-gcp-service-comparison`

- Azure for GCP Professionals: `https://docs.microsoft.com/en-us/azure/architecture/gcp-professional/`

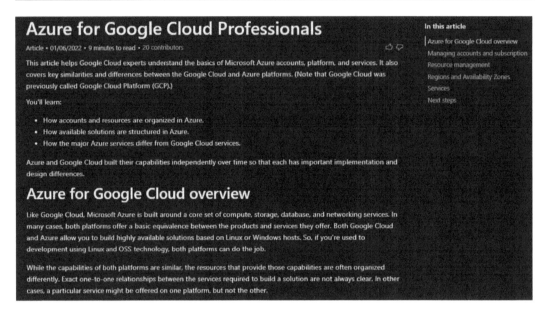

Figure 1.2 – Azure for GCP Professionals screenshot

- Azure for AWS Professionals: `https://docs.microsoft.com/en-us/azure/architecture/aws-professional/`

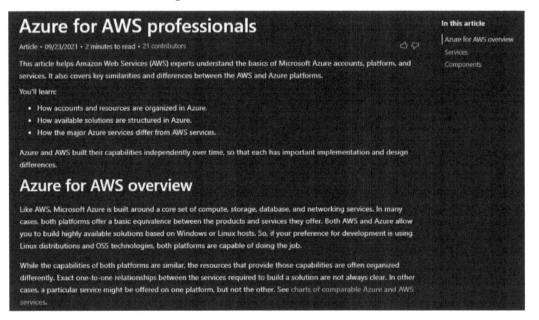

Figure 1.3 – Azure for AWS Professionals screenshot

Getting to grips with one cloud platform may seem like a daunting task. If so, you probably think that learning about all three is an impossibility. Rest assured that each cloud has many similarities and the skills you acquire now will stand you in good stead if you ever need to use another cloud in the future. Hopefully, these articles have enlightened you a little and shown just how similar the major cloud platforms really are.

Cloud workload types

A workload is a collection of assets that are deployed together to support a technology service or a business process – or both. Specifically, we are talking about things such as database migration, cloud-native applications, and so on.

When talking about cloud adoption, we are looking for an inventory of things that we will be deploying to the cloud, either directly or via migration.

You need to work across the organization with all the stakeholders to identify workloads and understand them, prioritize them, and understand their interdependencies to be able to properly plan and parallelize or serialize your workloads depending on their needs and dependencies.

You and the stakeholders will need to identify, explain, and document each workload in terms of its name, description, motivations, sponsors, units, parts of the organization they belong to, and so on. This then means you can further identify metrics of success for each workload and the impact this workload has on the business, on data, and on applications.

Then you can approach the technical details such as the adoption approach and pattern, criticality, and resulting SLAs, data classification, sources, and applicable regions. This will enable you to assign who will lead each workload and who can align and manage the assets and personnel required.

The highest priority must be given to a decision between migration as is (commonly known as *lift and shift*) or a more modern cloud-native approach. The highest priority must be given to this task as any error here will cause delays and, because of dependency issues, the timeline slip may escalate quickly. And with enterprise customers, there may be thousands of workloads to execute. Take care that this step is taken very seriously and meticulously by the organization.

One common thing that happens is that a lot of responsibility gets assigned to a very small team who may not have all the information and must hunt for the information in the organization while trying to prioritize and plan the workloads and dependencies. This usually results in poor decisions. While it might be tempting to go for modernization, where migration is concerned it is best to lift and shift first, followed quickly by an optimization phase. Business reasons for the migration are usually tied to contractual obligations (for example, a data center contract) and modernization for teams new to the cloud rarely goes swimmingly with a looming deadline.

On the topic of business cases for each workload, do remember to compare apples to apples and so compare the total cost of ownership for all options. This rarely gets done properly, if at all, especially if done internally without a cloud provider or consulting support.

Ensuring cost consciousness is another activity that gets overlooked. You need to plan before you start moving workloads around. Who will be responsible and how will we monitor costs. Overprovisioning happens often and with regularity. And remember: ensuring cost-optimized workloads is not a one-time activity. You are now signing up for continuous improvement over the lifetime of the workload, or risk costs spiraling out of control. Once they do, it is even harder to understand them and get them back under control.

Cloud pricing

As mentioned before, quite a few times now, agility not cost must be your primary goal. Having said that, letting costs spiral out of control is wasteful, so occasionally, (at least quarterly) every team should invest some time in optimizing costs. And if you are a great architect, you (or your team) will want to join in (or initiate things) and help out with synergies across teams they may have missed.

Counter-intuitively, cloud providers and their account teams should and usually are incentivized to help you optimize costs, so check in with them regularly if they don't proactively reach out to you. The reason is simple: the happier you are with the cloud performance cost-wise, the more you can do for the same amount of money, so you will do more in the future. It really is that simple.

OK, so you want to optimize your costs. What do you look at first?

The easiest is to start with two things: individual services and high availability requirements. Individual services are updating all the time, adding new cost tiers (for example, the Azure Storage archival tier), adding serverless options (for example, pay only for the actual usage on a per request basis, such as the Azure Cosmos DB serverless option), and moving features from higher tiers into lower ones, giving you the ability to trade off cost versus capacity, performance, and features.

The next best thing is that, thanks to overzealous reliability requirements at the start of any project, you can usually go back and architect around or remove completely such requirements and save considerably. For example, a calculation service that is deployed to 14 data centers because you started with one Azure region pair of 2 and then replicated that to all other paired regions and now have deployed that service 14 times because you have 7 two region pairs. Is this really required or could it be just 1 in each paired region and the fallback is to any of the other 6?! Beware of data residency requirements here, so maybe it still is a valid requirement.

Multi-regional failover is relatively easy and is often overlooked. With just a few DNS changes and a few Azure Traffic Manager settings, you can increase reliability significantly and quickly with little effort.

Other things you can do require a bit more effort, such as moving from one database type to another (for example, Azure SQL to Azure Cosmos DB), switching between comparable services, optimizing APIs to have them be less chatty, deploying to Linux machines instead of Windows, and so on.

> **SignalR**
>
> Sometimes you can get amazing results – for example, my favorite service in Azure is Azure SignalR, which is used to add real-time functionality to your apps. But if you think about it, real-time functionality is similar to querying a database directly, and if you have a lot of the same queries, there may be a way to use SignalR to execute the query once and have thousands of requests return the same response, like caching but not even having to query the cache, getting the response through a push mechanism before the request to then cache or database gets made.

Azure has a pricing calculator on the website, which you can use to get your overall estimate, but for cost optimization, it doesn't really help outside of showing you the reservation options. For example, if you have a standard baseline usage of some services (for example, Azure VMs, Azure Cosmos DB, Azure SQL, etc.), you may reserve capacity and prepay for it and get significant discounts – over 50% in some cases.

You will also get recommendations from the AI behind the Azure Advisor service, and while those are almost always great to act upon, quarterly reviews are still a necessity.

* Azure Pricing Calculator:

  `https://azure.microsoft.com/en-us/pricing/calculator/`

Cloud support

As for paid support, there are multiple options available. If you are playing around in a sandbox environment and you really don't need support, you will get some help when trying to make things work from Stack Overflow and other random blogs. However, only the official support can diagnose certain technical issues. Of course, in production, you will likely need a quick response time and help through service outages.

The support options in Azure are as follows (`https://azure.microsoft.com/en-us/support/plans/`):

Type	Description	SLA
Basic	Included for all customers, provides self-help resources	N/A
Developer	Access to Technical Support via email	Business hours
Standard	For production environments	24/7, 8-hour response
Professional Direct	Includes proactive guidance	24/7, 1-hour response
Enterprise	For support across the Microsoft suite of products, including Azure	24/7, 1-hour response

Table 1.1 – Compare Azure support plans

What is the Cloud Adoption Framework?

All the hyperscale cloud providers – all three of them (Azure, AWS, and GCP) – know that to get the most value out of your investment, you must adopt the cloud and the cloud concepts properly – otherwise, you will invest less in the future.

One struggle the account teams in these hyperscale cloud providers have (and yes, that includes the account team that works with you) is your speed of adoption, which is limited (and therefore impacts their KPIs and their promotions and bonuses) by your struggles with getting things done quickly, at scale, and with definitive and recognizable business benefits.

So, your and your organization's lack of proper cloud adoption is not only making it difficult for your organization to avail of the benefits of the cloud, but it may also in fact limit the cloud adoption by your internal teams. And of course, think of your cloud account team and their bonuses. While this is just a bit facetious, it really is not a win for anyone. Luckily, the only win case here is for everyone to win by adopting the cloud in the right way. Doing what this book advises is good for everyone, including the broader consumer market, and is the only way for your organization to stop struggling and start enjoying cloud adoption.

What exactly is the benefit of this book over comprehensive resources that are available freely online? Great question. What you will get from those guides is a lot of insight into the specifics of each cloud, but what you won't get is the years of experience working with clients, helping you avoid pitfalls and letting you know what and how to prioritize your way out of these. Also, all these lack any humor whatsoever. And sometimes they are just plain wrong – they lack any insight into your organization. Every organization has its quirks, its legacy issues, and its future plans, and so what we are doing in this book is guiding you on a path where you can confidently pick and choose (cherry-pick, if you will) what will and won't work for your organization.

Should you read all about every cloud service or just focus on the subset that your organization is adopting? Another great question, you absolute legend, but one you already know the answer to. Yes. Read (skim through) all the available documentation. You will learn a lot. Sometimes what you learn you will also remember if you've read it multiple times. You will also learn the subtle, nuanced differences between the providers. And you will learn what their priorities are and who their target audience seems to be. You might be surprised.

One thing you must resist though, is the temptation to adopt everything you read online, hence this book. Otherwise, you will sacrifice agility for premature optimization. And the providers' own account teams will try and take you on a journey of fully adopting these in the way they are written. This is 100% wrong. Calling it right here. Yes, absolutely you should work with them and their wealth of knowledge, but on your own terms after fully understanding the causes and effects each of the recommendations will have on one thing – your organization's agility (that is, your organization's ability to deliver business value).

So which hyperscale cloud provider is best?

Amazing question. So original. No, I always get asked it, having experience with all three clouds. So here is my definitive answer – just an opinion though, so think carefully before writing me a nasty "*Well, actually…*" note! It's a short opinion, so missing a lot of nuances, but you are not here for nuanced opinions – no one ever is.

Azure is best for two target audiences: enterprise companies and everyone who hates the AWS console. Enterprise companies cannot find a better partner out there than Microsoft. You are using Office 365 and/or you have legacy enterprise software and/or own data centers and/or need commercial support selling your software and services. No company other than Microsoft will serve you better or support you better. And Azure portal blades are the best thing since sliced bread. The AWS console is holding their customers back – literally. For start-ups, look elsewhere unless you are on the Microsoft stack, then pick Azure. However, you will be on your own – Microsoft will throw you a bone sometimes (such as through the Microsoft for Startups program), but it is up to you to get things done. Once you start scaling customers and profit, welcome – you are now an enterprise company. Talk to Microsoft again.

AWS is best if you are a start-up focused on business value rather than geeking out over technology. If technology is a means to a business end, AWS is for you. It has easily the best marketplace, easily the best support (hello, chat), and is the easiest path to take, if you are not all in on the Microsoft stack. All services are there for you. Just pick them and scale.

GCP is for the technology geeks and those start-ups with a deep affinity to the way Google services work. If technology *is* your business, GCP is for you. This is the true home of any SRE. And if you are in the advertising space, GCP is your valued partner. Do not buy into any early access or new and innovative market-making service though as Google is famous for killing or abandoning services. If all you do is AI, GCP is for you as well. If AI is a valuable piece of your overall business, you are better off with AWS or Azure.

Two final thoughts: one, you won't go wrong picking any of these if you are a capable individual and a robust and knowledgeable organization, so don't stress it too much; two, none of this matters anyway, as your organization's CEO will pick a *hyperscale* cloud provider, throw $50 or $500 million at them and commit your organization to them for the next 5 years (and beyond) and you will have to just deal with it. So there!

Summary

By now, you should have a good sense of what we hope to achieve with this book. You probably breezed through the cloud foundations, though you likely have many unanswered questions on cloud adoption. That is expected at this stage. We will hopefully answer these over the course of the book.

In the next chapter, we set the scene for every successful cloud adoption – strategy.

2
Adopting a Strategy for Success

With the introduction done, you are probably itching to dive straight in and start building your new home in the cloud. Whoa now, hold on! For those of you coming from a software engineering background, you know it pays to stand back and plan your approach before diving headfirst into the code. Don't underestimate the complexity of this transformation and know that it is more than just *IT* – you'll need to tackle the organizational culture and a few tech issues along the way.

Before you can get to a *plan*, such as a traditional IT project plan, you need a strategy – a plan of attack, a game plan, a playbook. Call it what you like but be sure to start here before pressing ahead.

In this chapter, we will highlight the most important part of any strategy, the goals, and the objectives you want to achieve. We will also review common cloud adoption scenarios that will help you frame your strategy.

In this chapter, we're going to cover the following main topics:

- Why do we need a strategy?
- Adoption scenarios
- What not to do

Why do we need a strategy?

Well, to start, any project needs (well, is usually given) a set of goals or objectives. At a minimum, this is what you need to achieve. This is what you, your team, and your entire cloud journey will be measured on. So, you had best be clear on what the goals are.

Do yourself a favor (you'll thank us) and aim for achievable and measurable goals. Don't try to disrupt the industry or revolutionize your client's experience – the cloud alone will not do that. Plus, how do you measure a revolution anyway? Usually, after the history books are written and frankly, you can't wait for that!

Goals that focus on cost, performance, reliability, and scalability are great – you can measure all these, though it's not very sexy. Cultural transformation goals, such as enabling greater agility, are trickier, so be very clear in the beginning about how you intend to measure it – define the right **key performance indicators** (**KPIs**) for all your goals.

Perhaps another piece of advice is to try not to have too many goals. You can't solve all the organization's problems, though you might think you can. The critical thing is solving exactly those problems and achieving exactly those goals you absolutely can and must for the success of the cloud adoption – you can always iterate to solve others as well, but taking on too much at any one time will cause KPIs and other measurements to be tainted by changing to too many variables at the same time. If you reorganize your organization, change the technology stack, use a different software methodology, and pivot to target a different set of customers – how exactly will you know which one of these changes was successful and which was not? *Experience*.

Once you go through one cloud adoption, then another, and then another, you will have the benefit of hindsight and then you may align multiple objectives at the same time to save effort and speed up the adoption. Even then, be careful not to be blindsided by your experience – different organizations, different markets, different technologies, and different customers may not react to the same set of changes in the same way. Things that worked before may not work anymore simply because of technology or market evolution, so always double-check your assumptions and rely on KPIs to drive you.

Key performance indicators

The right KPIs are those that align with the business goals, not just IT. Some teams, of course, will have their KPIs purely IT/technology-driven, but be extra careful that they are not opposed or contradict the business goals.

For example, a business may want to have a KPI around the number of customers onboarded, but if IT has the same goal, the customers may be onboarded to single-tenant instances because that is faster initially instead of a multitenant environment, which will cause some speed at the beginning but will result in huge technical debt further down the line. And in today's fast-moving world of technology and technology careers, the people taking on technical debt won't be the ones that will have to pay it off.

Define clear, concise, measurable KPIs and stick with them – up to and until you realize they are incentivizing bad behavior. Then, change them immediately based on the knowledge you have acquired. And trust that the KPIs should change often.

Even companies such as Microsoft do not shy away from changing KPIs yearly and modifying them throughout the year as well.

If you can tie KPIs to your customers' success, that's even better. For example, if you are a provider of instant messaging technology, a KPI may be the number of customers you've sold it to. A better one would be a percentage of people in your customers' organizations that use instant messaging technology, and an even better one would be a percentage of people in the customer's organization that are solving business challenges more easily through instant messaging. Can you think of an even better one?

How about business outcomes (in $$$s) that your technology has helped bring in for your customers? Can you think of an even better one?

Now, the questions you are probably asking yourself are – how can I even measure such a thing and how would that KPI look from an IT perspective? Measuring is hard, and measuring well is harder still, and measuring accurately and with future predictions baked in is impossible. Or is it? Yes, yes – it is – it is unbelievably hard. But would you rather measure an easy KPI that is next to useless or a KPI that drives transformation but might be a lot harder to measure?

The reason for this long rant is based on the experience of all the organizations I've worked with that have never measured KPIs or have measured KPIs completely misaligned with the business.

One final example I'll provide here is measuring agility. If you measure agility and you are delivering more and more features, services, or products, that may be the greatest thing ever – but it also may not. It is relatively easy to measure how many features are delivered, but measuring delivery of the "right thing" requires customer or user feedback.

There is a story, usually misattributed to Napoleon, German generals in 1933, and others. It talks about four officer characteristics – the clever, the lazy, the stupid, and the industrious. And to a casual observer – the clever and industrious are great, whereas the stupid and the lazy are not. However, this is not true. The clever and industrious can be made into high-staff appointees, while the stupid and lazy are great for boring work and as cannon fodder.

The officer that is clever and lazy is a great candidate for high command. But a combination of stupid and industrious is especially disadvantageous, to the point of the recommendation being to remove those individuals from the service – immediately – because they will do a lot because they are industrious, but because of their stupidity, they will be doing the wrong things – and lots of them. So, it's better to get rid of them immediately.

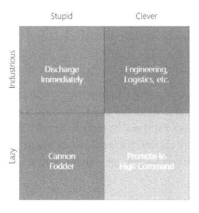

The same is true for KPIs.

Clever and lazy KPIs will target business-aligned goals but they will be difficult to measure and maybe even lag in measurement. This KPI will correctly guide you to your goal – for example, customer sentiment.

Clever and industrious KPIs will be live. Often updated, correct, and specific, these KPIs will help you guide your day-to-day and even minute-to-minute decisions – for example, the current volume of requests or the speed of database queries.

Stupid and industrious KPIs will be measuring wrong things very well; say, for example, the number of commits or velocity points, but may lead you to outcomes that, at best, can be described as random.

And stupid and lazy KPIs are better left to your immagination; surely you can think of some of them you or your organization might be using right now. These are the kinds of KPIs that both don't make sense to measure and are not easily measurable – for example, impact (search online for 'impact score' and see for yourself).

Goodheart's Law

There is an old saying that *you get what you measure* and another one that states *when a measure becomes a target, it ceases to be a good measure*. The second one has a name – Goodheart's Law.

> **Rat farms**
>
> This observation is from *Jingo*, a book by Terry Pratchett (my favorite book in the huge series of books set in the Discworld universe).
>
> It talks about how there was a rat infestation in Ankh-Morpork (a metropolis in this universe). The administration decided to include a 20-pence bounty on rats in the city. To qualify, you were to present the tail of a rat you caught/exterminated as proof.
>
> However, this did little to reduce the number of rats. Rather than encourage the city's residents to solve the rat infestation themselves as intended, it resulted in very little work being done and only incentivized everyone in the city to create efficient and very lucrative rat farms – free-market environmentalism indeed.

Goodheart's Law claims that all measures that become targets will be abused eventually, so keep that in mind when you're picking your measurements and KPIs. This is especially important in regulated markets and just in general in today's capitalist society with a lot of regulation on what is and isn't allowed – so not only can the poorly chosen KPIs lead your organization astray with regards to the business goals, but they can also land you in hot water with regulators and governments around the world.

Itemizing business outcomes

Sometimes, tech folks focus too much on the tech. Architects, of course, aim to connect the business with the tech, so it's crucial that the goals you establish have a clear path back to the *business goals* and that everyone (well, the stakeholders at least) understands how your transformation goals contribute to the desired business outcomes.

Typically, we often see a duality in business – the business and IT strategy. This makes little sense in practice since there is only *one* strategy. An effective architect must navigate this and ensure that the transformation goals are completely aligned with the business. This will make getting buy-in from stakeholders, team members, and others far easier. Also, this is the *only* way. If IT and business don't see eye to eye and don't have a complete understanding of what each is doing, then meaningful progress is very unlikely.

Transformation is hard and good; meaningful transformation is harder still. How can you transform into something if everyone in the organization is not 100% clear on the end goal? Everyone will be pulling in different directions, even if everyone has the team spirit to do well and do more. We cannot stress how much alignment with business is necessary. But having seen misalignment in many organizations, we want you to pause here and think – deeply. Are you aligned? Is everyone aligned? And if not – address that first.

Adoption scenarios

Now, let's address the different adoption scenarios. Not everything in this section will apply to you. While we encourage you to jump around this book and skip sections that may not be relevant to you or your organization, this section shows you all the different types of adoption scenarios based on different goals, so we want to encourage you to read this section in its entirety – and only at the end start thinking about your actual ones as they may overlap with these.

Cloud adoption scenario for a traditional enterprise

This is a view of a typical traditional enterprise. You know the type:

- Predominately internally focused IT that supports the business operations
- Investments are made over a medium to long term (5 – 15 years)
- Budgets are relatively fixed, with little room to maneuver when the needs change

There is likely already good news for you if the multiverse you are in is based on this type of scenario.

Primary (although, still not the only) motivations for this type of organization will include these things:

- Contracts (for example, data center contracts are expiring)
- World events (pandemic, working from home, supply line disruption, and so on)
- And very rarely, a true desire for modernization and the knowledge of how to go about it

Let's focus on contracts for a moment. Contracts with data center providers or just general data center amortization (data centers at the end of their life) are, nowadays, usually a moment when someone on the C-suite asks the dreaded question – what about the cloud? It makes sense to investigate the public cloud now that we are considering the future of our current data centers.

Convincing C-levels on the public cloud

This is very easy – in terms of swapping customers' own data centers for public cloud provider ones. I was part of an organization in Microsoft that could organize data center tours for customers – here's a hello to everyone in Microsoft's Customer Success teams. There are only a few regions in the world (out of 60+ regions) that cater to this. We planned to get the customer to give us a tour of their data center and then arrange a tour of one of ours (Microsoft's).

As well as being a fun bonding activity and showcasing Microsoft's commitment to sustainability, this allowed us to address any comparison questions between their data centers and Microsoft's up to and including questions of security – this was always a win-win scenario as it was crystal clear that Microsoft data centers are out of this world compared to their now considered *old* ones. A CEO or CTO or CISO would almost always say a line during the tour along the lines of – "I now feel embarrassed for taking you to our data center."

This was always just another touchpoint in addressing customers' concerns, but a great one in addressing issues in a fun setting outside of usual offices and Teams calls.

Establishing the total cost of ownership would usually be the approach. If this is done correctly, the cost always works out in the favor of the cloud for any large enterprise.

As mentioned previously, focusing on just the cost is very wrong. Agility in delivering business outcomes is what should be the primary motivation here. Cost savings can, should, and must come later but they must be secondary to agility; otherwise, your cloud journey will be a long and arduous one. And later usually means months or quarters away, not years or decades.

In the case of an organization such as this, what is important for you to understand is the actual deadlines on things such as data center leases and plan accordingly. In both cases, you either have plenty of time because the leases are 2 years out or you don't because the leases are 2 quarters out– the process must be the same.

To succeed, stakeholders must not only go along with the cloud adoption processes but they must take an active part in it and pull in the same direction with deliberate effort. Define your business outcomes and motivation for doing this, document them, and pull the trigger. In the case of large enterprises, you will almost always have the support of a team on the cloud provider side. Lean on them for any technical issues but take care that you and the teams in the organization own and understand the business outcomes for your organization.

Most of the effort of this kind of organization will be on lift-and-shift or pure migration, but also ensure that any low-hanging fruit such as like-for-like modernization is incorporated into the plan.

The desire (usually out of fear of the unknown) will be to do minimal modernization and just get things done with lift-and-shift. This will usually be a consensus decision and it will take courage to challenge it, but you must.

Here is why – in the vaguest wording possible. The cloud is great at doing cloud things, but to get the benefits of the cloud, you must do cloud things. Otherwise, you are using the cloud as just another data center and will get almost no benefits from the cloud-first principles – elasticity, scalability, efficiency, flexibility, reusability, interoperability, integration, observability, and so on.

An example of the pure lift-and-shift approach is VM migration. Moving hundreds of thousands of VMs with apps, databases, and storage might be easy but it is rarely the right approach. Yes, we do not want to overly complicate the migration, but some things should nevertheless be done.

Here are some concrete examples of modernization that you should always consider because they are easy to do:

- **SQL Server**: Azure SQL (caveat: there will always be a few exceptions, but in general, if you have 1,000 SQL databases, you may end up modernizing 950 of them while lifting and shifting 50; the same goes for the other scenarios)

- **Other databases** (SQL and NoSQL ones included): Their cloud equivalents such as CosmosDB or Azure Database for MySQL, and even things such as Azure Cache for Redis

- **Storage** (BLOB, files, and so on): Their cloud equivalents such as Azure Storage, Azure Files, and Azure Backup

- **Static websites from various stacks**: Azure Static Web Apps

There are many more examples, given how Azure has 200+ services ready for you.

Hopefully, you get the gist here – some of the services the public cloud provides are just so useful and simple to turn on that it makes sense to modernize immediately. For example, Azure SQL Server, which is a managed service, takes away a lot of the burden of managing a SQL database. Since you will be migrating the database anyway, migrate directly to this service if at all possible.

You now need to be armed with a total cost of ownership as-is with an inventory of the IT assets so that you can start prioritizing workloads to move.

There are two schools of thought here: start with the easiest or start with the hardest. So, here is a rule of thumb – always start with the easiest, except if the stakeholders still have objections to the cloud. In that case, start with a **proof of concept** (**PoC**) concerning one of the hardest workloads.

The reasoning behind both is that (almost) nothing is impossible to do with hyperscale cloud providers, so starting with the easiest gives you quick wins and builds your team's confidence and knowledge. And by the time you get to the complicated workloads, you will be much better equipped to deal with it, and it may be (relatively) easy as well. Only start with one of the harder workloads if you need to convince stakeholders. Do a PoC on the hard workflow and then start with the priority list from the easiest once you have started migrating.

Ensure you have plenty of resources before starting the migration so that it goes smoothly on that front – you don't want to be migrating while fighting for budget, people, and team time. That is a surefire way to do a poor migration and potentially kill cloud migration altogether – for now, of course, as the public cloud is not going away.

Ensure you have partner support – third parties, vendors, and others all play a role in your migration. Ensure they are with you and will support you. One example we see often is MFA appliance migration.

Regarding that – make sure that if you are dropping a partner or a vendor, they find out about it at the last moment. Remember the KPIs from earlier. They have them too, and often, they are not aligned with their customer – in this case, you. Be prepared to have a few tough conversations and legal teams ready to assist. This is a cynical statement but has been proven time and time again, so we would be amiss not to mention it. Sometimes, you might get very lucky and work with a partner and a vendor on a similar journey to the public cloud and can work together, but in all likelihood, they have already gone through this journey. Or you might get lucky in another way in that the partner or a vendor has a similar relationship with your cloud provider, so the cloud provider team should reach out to your partner or vendor to ensure a smooth transition. If you can read between the lines, this is another cynical one, but again, it's observed over and over again.

So, you have gathered the data and prioritized migration on a workload-by-workload basis. It's now the last chance to make a business case (if you haven't done this already), with the migration plan defined and scoped.

Remember how we said to start with the easiest workloads first? Here is another reason – skills. Skills across your organization may not be fully aligned with the work ahead of you, so start planning to upskill your teams. Start with the general cloud skills and then focus on the skills in the order you will need them – perhaps Azure networking and VMs and databases are more important than serverless Azure Functions.

You will also want to standardize every deployment and every workload rather than have individual approaches for each one. Standardization is so close to automation. And both have so many benefits, such as being less error-prone and less dependent on heroes and heroic effort.

Before you deploy any workloads, you have to have a place to land them at. So, let's call that place a landing zone. This will contain all the shared services such as networking, storage, security and observability tools, and so on. You will need to plan and build the landing zone. Luckily, all hyperscale cloud providers have a detailed landing zone plan that will help you narrow your work down. Unfortunately, being an enterprise organization, you will almost certainly require connectivity to and from the public cloud. This will complicate and lengthen the migration start by months or quarters (as third-party vendors such as ISPs need to get involved), so plan for that.

Assuming you have deployed and are happy with your landing zone and have a team that understands the landing zone and where different workloads will land, you are good to start implementing the 10-10 parts of the plan. We want you to migrate the first 10 easiest workloads. Take your time, learn what works and what doesn't, and make adjustments to the teams and timelines. Once done, continue until you migrate the first complete 10% of workloads. By the time that is done, there will be very little you won't be able to do in the cloud already – with the services used in that 10%. Hopefully, there was some modernization and not just a lift-and-shift effort.

After/during the first 10 workloads, they will need to be handed off to the cloud governance and operations teams. The migration team focuses on migration and governance, while the operations team focuses on managing them once they've been migrated. A great handover is key here; otherwise, the migration team will keep getting pulled back to manage the migrated workloads. The good news is that this minor conflict in handover will ensure it quickly improves over time – by all teams. Might the smoothness of the handover be a KPI? Might you measure how many times the operations team needs to contact the migration team after than handover? Yes, and yes.

Don't forget to communicate the work you've done with the organization and celebrate your successes. This seems nice to have but it is essential to ensure that everyone in the organization can see the steps toward the inevitable cloud adoption and maybe get them more interested in helping as well.

If your 10-10 plan didn't uncover major issues that required you to stop the migration, you are well on your way to continuing. Celebrate every time you reach another round percentage, and beware of the 80-20 rule – 20% of workloads will require 80% of effort – and in this case, usually, the last 20% will be the hardest as those will be workloads that are missing knowledge and ownership or that have a lot of technical debt. Maybe consider bringing in a partner/contractor with specific skills to be able to get those workloads done faster. Along the way, you may discover additional low-hanging fruit for your modernization efforts, so go for it if it makes sense and doesn't change the timelines by much or at all. You will also discover that there are a non-insignificant number of workloads that nobody needs, so take some time to plan to kill them instead of migrating them.

The 80-20 rule may be more of a 90-10 or 99-1 rule for your particular organization, so beware and plan ahead.

Cloud adoption scenario for a new start-up

This is a view for a new start-up. You know the type:

- IT is the business and supporting customers is what matters; supporting the business is secondary
- Greenfield, very little to migrate
- Money may be tight, people are in multiple roles, and prioritization is everything
- Day-to-day-based; can refocus or pivot quickly

The view of a start-up is significantly different than the view of an enterprise organization, namely because there is a lot less legacy code and technical debt, and therefore a lot less focus on migrating existing assets and a lot more on software development, feature deployment, and agility and the ability to pivot from one use case to another, if necessary, when trying to determine a product-market fit.

Focusing on your customers is key. Building in a vacuum is wrong. Getting feedback early and often is valuable. Delivering value to customers quickly or failing to do so may make or break your organization. These should be your priorities. After getting that sorted, your next (parallel) task is to have the technology to deliver said value.

Here, the focus is on agility – the speed with which you deliver value to customers. Luckily, cloud adoption can help you achieve this. **Platform as a Service (PaaS)** from Azure provides fully-managed services that allow you to focus more on achieving business objectives and less on deploying, securing, and scaling complex software systems. Here are some examples:

- Azure App Service for hosting your applications rather than deploying them on VMs

- Azure Kubernetes Service rather than managing your own K8s clusters directly

- Azure Functions for running arbitrary code, focusing on writing code, and event-driven execution, rather than managing VMs and their environments

- Azure SQL and Azure Cosmos DB to (somewhat) abstract away the complexity of high availability and disaster recovery – you still have to know and plan them but there are tools built into the service to assist you in managing them

Then, there are services you need that would require specific knowledge, staff, or investment to do well. These are not (usually) a part of your core business focus:

- Azure DevOps to plan and manage your tasks and environments and deployments through **continuous integration/continuous delivery (CI/CD)** pipelines

- Azure Active Directory to manage your users and things such as single sign-on

- Microsoft Defender for Cloud to protect your environments in Azure and elsewhere

- Azure Front Door for web content caching, application firewalls, domain management, and much more

And then, depending on your core business, some services might be ideally suitable for your business:

- Azure SignalR for adding real-time functionality to your services

- Azure Maps for creating location-aware services of your own

- Azure Cognitive Services so that you can avail of cognitive capabilities (focused on speech, language, vision, decision, and more) as easily-called API endpoints

- Azure Digital Twins for creating models of physical environments

- Azure Orbital for fast linking data between satellites and ground stations

As there are well over 200 Azure services, you might invest some effort into evaluating their suitability for your core business as they may give you a significant head start advantage if they align with your business.

The best cloud concept you must familiarize yourself with is **elasticity**. This is another crucial thing when bootstrapping your business. Elasticity is, in essence, a way for you and your organization to pay exactly for the services you use when you use them, provision more only when you need to, and provision less (deprovision) when the traffic from your customers lowers – this may be daily (for example, during the night), weekly (for example, during the weekend or on Friday afternoon), or monthly (for example, in January).

The focus of your engineering teams should be split between working on services and features you want to deliver to your customers and managing the platform you've built in Azure. The emphasis should be much more on the services and features as opposed to managing the platform – hence, we recommend using PaaS services whenever possible.

Even though you are managing a start-up here and maybe struggling financially (at some points in your journey at least), you should still be focusing your efforts on agility and getting those features/services out there into the hands of your customers, and only then on cost savings. This agility will let you quickly determine if you need to adapt your features/services, double down on them, or discontinue them. This will then enable you to focus on cost optimization with the right mindset. That doesn't mean breaking the bank, but if the choice is between shipping now or in a few weeks/months, then ship now and optimize in a few weeks/months. If you are smart about your selection of PaaS services, you will be fairly safe in terms of unexpected costs. This leads us to **governance**.

You can accomplish a lot of the Azure governance using the Azure Policy service. If you have engineering teams in Europe, you can restrict all dev deployments to European regions. You can also restrict services that are and are not allowed to be used in Azure – for example, you can restrict the deployments of VMs.

Building a consensus around the policies, services, and technologies, in general, should be easier in a start-up environment. However, some thought should be put into this, primarily around the risk register and the technical debt registers – you will need to revisit these, so you might as well document them now. This will save you a lot of unnecessary problems down the line.

> **Microservices or a monolith?**
>
> You will get a lot of advice on building anything, including a monolith at this stage, with the hopes of proving the business value and market fit first; then, you can optimize your technology choices.
>
> This is the worst idea ever, and usually, I've seen it from non-technical founders or investors. While it does make sense at a PoC stage to just build something ugly fast, there is very little benefit of this approach for production uses.
>
> Building a microservices structure for your services isn't that much harder than building a monolith. It may be hard for someone coming from enterprise IT (nowadays, I doubt even that), but there are so many organizations availing of the cloud and so many start-ups doing the same that if you and the people in your current organization are new to this approach, it should be relatively easy to hire someone that has been through this before, even if it is a temporary advisory role initially.
>
> Save yourself considerable effort and pain and start properly. You are building your services so that they can scale, right? Otherwise, what is the point of them? Or what is the point of your start-up?

Yes, sometimes a monolith makes more sense than microservices, and sometimes it does not. You need a lot of experience to make that judgment and a lot of knowledge about your organization and its objectives. The good news is that of course you can change your mind and rewrite/rearchitect/refactor, and the bad news is – it takes effort to do it.

Cloud adoption scenario for running a modern SaaS platform

This is a view for running a modern SaaS platform (such as Spotify, Netflix, GitHub, Zoom, Salesforce, LinkedIn, Revolut, or YouTube). You know the type:

- Can be a startup business scaling quickly or a unit of a larger enterprise.
- Global reach to customers in all markets.
- A monolith that doesn't scale, or thousands of microservices that are unmanageable since they've been developed quickly and without an overall strategy.
- Likely multi-cloud due to acquisitions.
- With an idea for a portfolio. At the moment, the portfolio is emergent (we have what we have as that is what we've built) as opposed to planned (we built it with purpose and the services fit well within their domains and each other).

Contrasting this with the previous section that dealt with start-ups in their relatively early stages, this section is for start-ups that have a global presence or for a unit of a larger enterprise that has public-facing services it is delivering globally.

The good news is that you are already blessed with customers, a lot of them, and they are (more or less) distributed around the globe. Congrats on that.

A typical issue we've seen at this stage is that of **scale**. You may have built some services that are now getting used more and more, but how you've built them may not be the best. It is now time to bite that bullet and plan for paying off technical debt. This absolutely won't get done unless it is prioritized ahead of other work, so you may be the voice of reason that must push for this. The teams will almost always be aware of the issues already and will wholeheartedly agree this needs to be done. The symptoms they will see are as follows:

- Not enough automatic testing, so manual tests are needed to verify functionality or bugs

- The time to implement new features is lengthening and/or implementing any new features means touching many layers in the technology stack

- Onboarding new team members isn't easy as it involves a lot of hand-holding and up-front explanations, especially concerning why something has been done (from a historical perspective)

- Customers complain their desired features are not being delivered as promised – usually due to the team velocity decreasing

- A team exclusively working on the implementation of features or services that does not benefit the customers immediately

So, do yourself and the organization a favor and advocate for paying off some (or all) of this technical debt now. Managing your own SaaS environment is hard, but it is an end goal for ultimate scale, flexibility, and, ultimately, agility and simplicity. The end goals are as follows:

- To have a cloud-native, cloud-hosted application (services) so that you don't have to manage infrastructure or, even worse, data centers

- To have a marketplace of integrations so that your services work for as many people and as many scenarios as possible

- To have flexibility in terms of new features and services from a business standpoint and flexibility in terms of technology and templates used to develop and deploy new services – new use cases may be discovered that increase your market share significantly

- To have an easy-to-understand (note we haven't used the word "simple") pricing calculator so that, internally and externally, everyone is clear on the cost of deployed and utilized services

- To have a hassle-free implementation for first (your engineering teams) or third parties (other developers)

- To have a self-sign-up to simplify onboarding new customers – at scale, you can't be handholding anyone, but your support and portal materials better be of the highest quality

To prevent facetious feedback on this part, yes – it is true that maybe in your particular case, one or two of these goals might not be applicable, but you better have a very good justification before you discount them.

Your focus on this will be mostly around building a consensus and then a plan on how best to achieve scale, getting the buy-in at all levels of the organization, and either bringing in the skills necessary to accomplish this or leveling up the existing skills in your organization.

What not to do

We've shown what *is* to be done, so let's see what shouldn't or mustn't be done. These are the common pitfalls and anti-patterns – the things that may seem obvious and perfectly acceptable but that nevertheless aren't.

Here are some anti-patterns to avoid (in random order):

- Accumulating technical debt without justification and a plan to address it – this is never a good idea anyway, but it is especially bad if the cloud adoption is trying to address and improve the time-to-market. Remember, if you are building it or you've built it and it isn't in the hands of your customers, what good did any of that effort do?

- Lift-and-shift with no modernization plans – the cloud wasn't meant for this, except in a very narrow case of maintenance and no further development until the service gets decommissioned. That is the only acceptable case of lift-and-shift.

- Assessing the skills available in the organization, finding them lacking, and then not immediately addressing them – you either upskill, hire, or get partners to support you; otherwise, you fail.

- Not getting the buy-in for cloud adoption at the highest levels – and then constantly and tirelessly evangelizing the cloud every step of the way throughout the entire organization.

- Thinking that the cloud is just another data center or thinking that the cloud is business as usual – from both business and technical perspectives. The cloud is a completely different beast, a completely different set of rules and expectations.

- Over-engineering reliability, high availability, and disaster recovery.

Over-engineering

I've seen this so many times. Cloud services are meant to fail and fail quickly and gracefully. They are not generally meant to be so reliable that your focus and your cost are skewed toward them. Everything except your core business services must be able to fail and fail often and fail at random times.

If you are running a streaming service, everything except browsing for and streaming content is superfluous (and accepting payments, of course). If you are running a bank, everything other than making and receiving payments (yes, even inter-banking) is superfluous. And so on.

Yes, it's great when all the services work all the time, but between some or none of your services working, you better focus on keeping your core business services up. All other services are there to support your core business. Should you make every effort to maintain them? – Yes. But if you must make determinations on what to support and what to prioritize in an emergency – some services are more valuable than others.

Similarly, while the storage is cheap (relatively) there is no reason to have backups replicated 10+ times, in multiple regions, in multiple clouds, and on-premises. Enough is enough. You can have as many as you need, but not more.

High availability for the services that do require it should be baked in and not rely on a failover mechanism that will see its first test (admit it) when it needs to be used for the first time.

I've seen these edge cases often and not only do they cost a lot to implement, but they also complicate deployments, backups, recoveries, and failovers quite a bit. A simple, straightforward plan that defines RPOs and RTOs properly is sufficient. And before you define them at a technical level, talk to the business – you will uncover many SLAs that are too high due to assumptions on the business case being critical. Every single time I've reviewed RTOs, RPOs, and SLAs, I've always discovered overprovisioned infrastructure to support them – unnecessarily.

- Diving into the migration without a defined plan, structure, landing zone, and KPIs. So, you have migrated some or all workloads. Are they now performing better than before, are they more or less costly, and are they a help or a hindrance to your development and operations team? How do you know if you don't set up a proper plan and execute it, and then adjust it to address gaps?

- Forgetting about shared responsibilities. While it is true that the cloud provider will do a lot to improve the platform's security, ease governance, and decrease cost, it is up to you to share that responsibility as well. You can use secured services and be insecure, you can govern your cloud in suboptimal ways – even with the tools provided by the provider – and you can run up costs even though individual services are cheap, including the cloud provided tools and AI-based reports that help keep costs low.

- Failing to monitor performance. In the cloud, you are using commodity hardware and shared infrastructure (with other customers) on a common network – the internet. All these and more mean that monitoring performance is essential as it will vary wildly in terms of hardware behavior, noisy neighbors, and global internet traffic and routing.

- Assuming current IT staff can just jump on over and assume the same or similar roles in the cloud. The cloud is radically different and different skills apply. Experience plays a crucial role as well. Only the right combination of skills and experience can make you successful. Forget about design by committee and allow people with skills and experience to shine and lead the way forward; otherwise, you will repeat the same mistakes. Sometimes, these mistakes can be fatal – for your career or the life cycle of your business.

Summary

In this chapter, we looked at various cloud adoption scenarios, what needs to be done, and what you should avoid. One common theme throughout was that starting with clear, measurable goals is critical for success.

In the next chapter, we will explore the Cloud Adoption Framework, why it exists, and how to use it, and build a plan to help you guide your journey to the cloud.

Part 2: The Plan

Having the ambition, strategy, and goals for cloud adoption is a great start. Now you need a plan to execute. The planning phase is crucial for success. Timelines, budgets, stakeholder alignment, and risk management are important. But so too are the early technology decisions, impacts on ways of working, and wider people organization. The planning phase is complicated as there is no single solution or best practice to follow, but there is a framework that can help. In this part of the book, we will introduce the Azure Cloud Adoption Framework and explore the many challenges and benefits of cloud adoption.

This part of the book comprises the following chapters:

- *Chapter 3, Framing Your Adoption*
- *Chapter 4, Migrating Workloads*
- *Chapter 5, Becoming Cloud Native*
- *Chapter 6, Transforming Your Organization*

3
Framing Your Adoption

In the last chapter, we discussed why strategy is important and looked at common adoption scenarios to help you define an approach that works for your business. But strategy without execution is just a bunch of ideas – exciting ideas with amazing potential, sure, that are great to talk about with your colleagues, but just ideas, nonetheless.

Now, the work must start, but how can you turn these great ideas into reality? Once you start to pick apart each idea and discover what actually needs to be done, it can become overwhelming very quickly. You need to be ready for cloud adoption, meaning your technology and people are ready. You also need to know what you're getting into and have a plan. If you are new to the cloud, a lack of experience can prevent you from creating a credible plan – you don't know what you don't know. Even if you have many years of experience working in a cloud environment, and you feel confident that you have mastered the technical concerns, what about business alignment? Change is hard. Any meaningful transformation requires many people to contribute – you can't do it all alone. And what happens after you have successfully migrated to the cloud? Is it ever really done?

What you need is a way to break down the adoption into discrete parts, to put some basic structure in place. A framework will help you organize the complexity and make the adoption appear less overwhelming. Luckily, such frameworks already exist! In fact, all the major cloud providers understand the complexity facing businesses and have developed frameworks to help you adopt the cloud. Of course, it is in their interest to do so; more successful adoptions equal more recurring revenue for them – which is a win-win – if you are successful in your adoption.

Microsoft's **Cloud Adoption Framework** (**CAF**) provides tools and guidance to help you plan, design, migrate, and operate your cloud environment. CAF helps to streamline the process of adoption with best practices, checklists, templates, and assessment tools.

In this chapter, we're going to cover the following main topics:

- CAF – we will introduce Microsoft's framework for Azure
- The adoption journey – describing what the road ahead looks like

Cloud Adoption Framework

The Microsoft CAF aims to help you achieve your business goals by providing guidance, best practices, and lessons learned. This is a big-picture view of adoption, rather than narrowly focusing on specific technologies or services.

(The original diagram can be found here – `https://docs.microsoft.com/en-us/azure/cloud-adoption-framework/_images/caf-overview-new.png`):

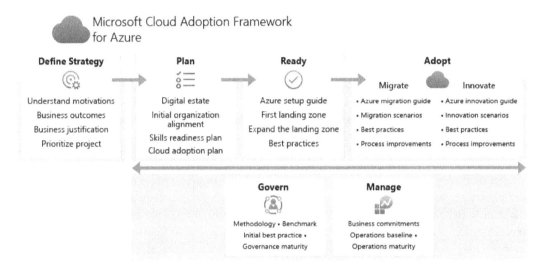

Figure 3.1 – CAF phases

Before we dive into all of this, however, it is necessary to understand the CAF and what problem it solves. The CAF helps to structure people, processes, and technology within an organization in order to make any cloud adoption successful. It provides guidance on how to establish governance policies, controls such as cost optimization, security baselines, identity management, resource consistency, and deployment acceleration in order to ensure a successful cloud adoption.

The CAF is organized around six phases – Define Strategy, Plan, Ready, Adopt, Govern, and Manage. Within each phase, there are discrete activities that produce some output required for a subsequent phase – kind of like a waterfall. But do not be led to believe the framework imposes any project management methodology. The cloud promotes and enables agility; your adoption should take advantage of this.

The CAF provides a comprehensive set of adoption tools and resources to help you make your cloud journey successful. These include the cloud adoption strategy evaluator, cloud adoption plan generator, readiness checklist, CAF foundation blueprint, and migration templates.

In the following sections, we will discuss how to use the CAF to prepare and embark on your cloud journey. We will also provide a walk-through of each phase, as well as discussing buy-in from the organization at large and skills an architect should already have before embarking on their cloud adoption journey.

Strategy

"Not again!", I hear you scream. Sorry, but we need to reinforce this point: it is vitally important to know what you want, why you want it, and how you will get it – or at least, at this stage, know how to measure success. It is so important that the whole topic has a dedicated phase in the CAF.

Strategy is about deciding what you want to achieve and then working out how best to do it. Sometimes it is called a "grand strategy" because it sets out the high-level direction for an organization, its overall aims, and how it will achieve them. It can be summarized in a mission statement or vision statement.

A cloud adoption strategy needs to align with the expected business outcomes (i.e., goals and objectives). But what makes a good strategy? There are a few key things to consider when defining your strategy:

- A clear definition of what cloud usage means for your organization

- A phased approach to adoption for minimal disruption

- The use of CAF resources and templates

- A focus on key workloads and applications that provide the highest value

- Regular review and measurement against objectives to ensure success

A cloud strategy for a business is the process of developing a plan for utilizing cloud computing technology to achieve specific goals. The goals may include reducing costs, improving efficiency, or increasing agility. Achieving these goals requires a clear understanding of what cloud computing can offer and a well-crafted plan for adopting it in a way that aligns with the company's needs.

The CAF cloud adoption strategy evaluator tool is a great way to assess the business readiness and identify gaps in thinking (or misunderstandings) and could help highlight new opportunities to justify your cloud adoption journey. The Strategic Migration Assessment and Readiness tool should be used while developing a cloud strategy and revisited multiple times through the adoption phases, to help you review where your organization stands in terms of cloud strategy, skills, usage, and technology capabilities.

The CAF provides a Strategy and Plan template (Word) to help you define a clear cloud adoption strategy to share with and get alignment from other leaders and stakeholders. The following sections we will help you with completing the main sections of your strategy document.

Understand the motivations

Firstly, you must understand what drives you to adopt the cloud. There are many reasons businesses invest in the cloud. Some of the most common reasons include reducing costs, improving efficiency, increasing agility, and achieving other business goals. Cloud computing offers several benefits that can help businesses achieve these goals. For example, the cloud offers economies of scale, which can help businesses reduce costs. Cloud services can also improve efficiency by automating tasks and enabling employees to work from anywhere. Cloud services can increase agility by allowing businesses to quickly deploy new applications and services – faster and without, in a lot of cases, a major upfront investment.

There are many potential motivators for businesses to adopt the cloud, but the CAF identifies two key triggers: innovation and migration. Innovation is about transforming products and services, gaining global scale, and increasing agility. Migration is typically the process of moving existing workloads to the cloud to optimize costs and reduce operational responsibility.

> **Vendor swap**
>
> I have heard many business folk (i.e., nice people who have zero interest in IT) over the years describe cloud migration as just **IT outsourcing** – it's just a simple vendor swap, a changing of the guard if you will. While there might be a sliver of truth in the statement, it is grossly misleading. Firstly, no IT vendor replacement project has ever been simple. Nor has any cloud migration – not when you could innovate while you go. Combining migration and innovation can be risky and difficult to measure. Don't try and make too many changes at once.

Business outcomes

This is where our strategic goals come in. Define specific measurable goals that are business focused. Some possible business outcomes of adopting the cloud include the following:

- **Reduced costs**: Cloud migration can lead to significant reductions in operating costs, as organizations can move away from **capital expenses (CapEx)** to **operational expenses (OpEx)**. An example goal might be to decrease the operating costs of the HR application by 10% within the next year.

- **Enhanced security**: Cloud providers frequently invest more in security than individual businesses, leading to a safer environment for data and applications. In business terms, the goal might be to reduce costs or the time taken to secure, audit, and certify the IT workload.

- **Global scalability**: Cloud providers have massive global infrastructure that can support an organization's growing needs. An example goal might be to increase the number of customers using our service within APAC by 25%.

- **Increased agility**: Cloud environments can be quickly provisioned and scaled up or down as needed, allowing businesses to react quickly to changing demands. In business terms, the goal is likely to reduce the time (and thus the cost) it takes to new IT resources where they are needed. Reducing red tape around IT procurements can significantly change the business culture and foster greater innovation. It is worth measuring this goal as there is always the risk that the procurement process remains the same even after cloud migration.

Business justification

Most businesses require some justification for spending money on anything. Unfortunately, cloud adoption is no exception. As an architect, you will be familiar with the Azure pricing tools, but estimating costs is just a minor part of building a business case. Unless you are also an expert on the commercial side of the business, you will collaborate with finance, sales, IT, and other departments to create a financial model or business case to support the expense and forecasted return from delivering the business outcomes. The business case should include the following:

- The estimated costs and savings benefits from cloud adoption
- A budget for partners to help with the adoption by providing skills and expertise
- A plan for measuring progress against the adoption goals and objectives
- A high-level timeline of cloud adoption phases
- An early outline of any organizational changes that are required as part of the adoption
- A clear exit strategy in case cloud adoption is not successful

> **We are all in operations**
>
> The finance team/CFO is your main ally here. They typically understand the cloud economic model, and to them it sounds like a no-brainer. Whatever you do, though, don't burn them. Adopting the cloud can lead to a lack of transparency, specifically on cost controls. A finance team that receives ever-increasing monthly bills from Azure without any clear explanation is not a happy team. We would recommend filling any cloud knowledge gaps in the finance team as a priority. The Azure learning paths are well thought out and could encourage the organization to adopt a *FinOps* culture. Eventually, I predict there will be an *EverythingOps*, or maybe we'll just go back to calling it Ops… Ah, the old reliable IT (r)evolution.

Prioritize the project

The CAF recommends that you avoid a big-bang approach, that is, trying to deliver all the business outcomes completely at once. Let's be honest, things will go wrong. Would you like to upset all users or contain the impact to a willing and very understanding select group of users? Do you want to manage complex interdependencies between multiple migration teams? How about achieving 20% cost reduction in six months, rather than 80% in five years? Keep it simple if you can, especially if it's your first time working on a cloud migration.

Consider the motivations for cloud adoption. If the aim is to migrate applications to the cloud, it may be possible to group applications along specific business functions or application types. Migrating a small, low-risk application group may be a good place to start for adoption. This way, you can achieve some success and learn from the experience before tackling larger or more complex migrations.

Now, some guidance on different ways to prioritize adoption goals:

- **Importance**: Some goals are more important to the business than others, so they should be given a higher priority

- **Urgency**: Some goals need to be achieved sooner than others and could have an opportunity cost associated with delay, so they should be given a higher priority

- **Cost**: Some goals are more costly than others, so they should perhaps be given a lower priority

- **Risk**: Some goals are riskier than others, either involving significant technical complexity or business critical applications or requiring resources that may not become available in time, so it may be prudent to give these a lower priority

Plan

The next phase is about planning the technological and cultural changes that need to happen. The Plan phase is focused on assessing your technology and people readiness for the cloud, identifying the known unknowns (gaps), and thus beginning to understand how disruptive the cloud adoption will be. Unfortunately, there is no single plan that fits all scenarios, but the CAF provides guidance and tools to help.

Once you have your cloud adoption strategy defined, the cloud adoption plan can be developed. Here, you should outline your cloud target state and define the teams and timelines and refine budgets for cloud adoption. With this plan in hand, you will also need to review technology and people readiness and identify any gaps that exist between the current IT environment and the cloud target state.

There are a few common types of plans that you might need during cloud adoption:

- **Cloud adoption plan**: This plan describes how the organization will adopt cloud technologies and practices. It includes the steps required to enable the cloud on each process, the resources needed, and the timeline for completing cloud adoption.

- **Cloud migration plan**: This plan describes how the organization will migrate its applications and data to the cloud. It includes the steps required to migrate each application, the resources needed, and the timeline for completing the migration.

- **Cloud governance plan**: This plan describes how the organization will manage its cloud environment. It includes the policies and procedures for governing cloud resources and the roles and responsibilities of cloud stakeholders, as well as how compliance will be ensured.

In the following sections, we will explore three key inputs for your cloud adoption plan, namely, understanding the current state, how the organization might need to change, and what skills will be required and how you will acquire them.

Digital estate

Here, we consider the IT assets of the organization or a subset of them potentially affected by the adoption project. You must carefully document these assets by creating an inventory or registry of applications, and for each, describe the infrastructure that supports it, identifying the business purpose and assigning an owner for the application. The more information that can be captured and centralized in your asset registry, the better. It is likely people working on the adoption will need to refer to and maintain the registry at multiple stages. Depending on the IT maturity, particularly within regulated businesses, an asset registry is likely to already exist to satisfy compliance audits, for example.

The CAF recommends labeling each asset to record the planned target state for the asset on the cloud. The five Rs of rationalization is one way to label them:

- **Rehost**: A "lift-and-shift" approach where no significant change is made to the architecture. Cloud services such as virtual machines are utilized to host the application and data instead of physical servers, but the application software stays the same. This allows the business to benefit from saving on CapEx, for example, where application software changes are too costly or impossible.

- **Refactor**: Typically, this approach considers minor changes to make better use of PaaS. Cloud services emulate some aspects of your current production environment, allowing workloads to be migrated, provided you can make changes to integration points, for example, changing the application's database driver so it can connect to a PaaS database server, or adding some JavaScript to the frontend UI so that errors can be traced within Azure Application Insights.

- **Rearchitect**: A more invasive approach than refactoring that aims to address any system incompatibilities with the cloud. This is a good choice if the system has evolved over time in an ad hoc fashion and now resembles "a big ball of mud" – that is, it has no obvious architecture. Systems with many software components, databases, file servers, routers, proxies, and "essential" jobs (either manual or scripted) to hold it all together deserve the attention of an architect.

- **Rebuild**: This literally means recreating the application so that it is fully cloud native. Cloud services and technologies such as containers, serverless functions, and microservices can be used to build the application from scratch. This allows the business to gain more benefits from the cloud and discontinue the use of existing legacy systems that no longer meet the needs of the business.

- **Replace**: Where possible, if a SaaS option exists, then the asset could be retired. A cloud service could supplement or replace the capabilities of your current production environment, while reducing or removing vendor lock-in.

Create harmony

In some scenarios, rationalization can be a dirty word. Behind the vast majority of (if not all) business-critical systems are people who care for these systems and are proud of their contribution to the business success. They know all the system faults, inefficiencies, and workarounds only too well. They are happy to share their war stories about 3 a.m. firefighting and turning around major upgrade failures. They may seem enthusiastic about adopting the cloud, on the outside at least. I like the five Rs of rationalization because they constrain the possible options, are easy to communicate, and make it easier to make decisions. But be mindful that some folks may convince themselves that what you really intend to do is down-size, retire, or make their role redundant! You may want to use softer language. Try the phrase "cloud harmonization"; it's worked for me.

Initial organization alignment

A stakeholder map is an important artifact. Understanding who the decision-makers are, along with those who are affected, helps to plan activities that support alignment. This is the softer side of the adoption. To successfully adopt the cloud, you must win the hearts and minds of those in your organization. Any veteran architect knows this is easier said than done. Adopting the cloud is likely to change the way people interact with systems and change the way they get their work done. The reaction to this change will range from indifference to blatant resistance. It is crucial that there is buy-in and alignment to what should be shared goals. This will likely require many meetings and many iterations of "the message" and possibly different variations tailored to specific groups so that it resonates best with your stakeholders and the people most impacted by the cloud adoption.

Skills readiness

A detailed analysis of existing roles and skills within the organization can help to identify resources for the adoption plan. But what about after the migration? There are likely cloud skill gaps in operations, finance, software engineering, IT, and client support – depending on the business environment you find yourself in. IT teams absolutely need to acquire new skills and experience. For example, the cloud can be a very secure environment, but this does not mean there is no risk. Cloud providers have rigorous security practices in place and are constantly working to improve them. But as with many systems, the cloud is only as good as the people who manage it. Just knowing what threat vectors exist on the cloud could be a significant knowledge gap, let alone having the skills required to define and implement data protection policies and procedures. But understanding how large the skills gap is helps to create a more credible adoption plan overall.

A cloud skills readiness assessment is an important part of any cloud adoption. It identifies the gaps in skills and knowledge that need to be filled in order to successfully adopt the cloud. The skills readiness plan should outline the steps that need to be taken to fill these gaps, such as training, hiring new staff, or partnering with an experienced service provider.

Cloud adoption plan

At some point along your adoption journey, someone is going to ask that dreaded question: when will it be done? Most likely, this will be asked many times by different stakeholders. If you feel this is a simple question to answer, pause and consider for a moment the Dunning-Kruger effect. This is a psychological phenomenon in which people with little knowledge or skill in a subject overestimate their own abilities, while those who know more about the topic often underestimate theirs.

The Dunning-Kruger effect can be very dangerous in a business setting. When people who are not skilled in a subject overestimate their abilities, they can make bad decisions that can lead to disaster. For example, an inexperienced manager might make the decision to move to the cloud without understanding the risks involved. This could lead to data loss, security breaches, overspending, or other problems. Creating a plan is probably one of the hardest parts of cloud adoption, at least for first-timers.

> **Someone needs to be in charge**
>
> While you want to bring everybody along on the journey, have all the input and understand everyone and everything in and about your organization, do not support design by committee – that won't work; it never works. Folks that know what they are doing need to be leading this transformation.

While the plan should focus on relatively small but achievable deliverables, there can be some inescapable tasks depending on the size of the business and the business operating environment. There could be significant regulatory considerations that require policies to be updated, compliance evidence to be recorded, and audits to be scheduled. Perhaps the transition to the cloud will impact the existing business client, at least contractually, and so clients will need to be engaged.

A cloud adoption plan should be aligned with the business strategy. The plan should include a description of the business outcomes that will be delivered by moving to the cloud, as well as a description of how the cloud will be used to achieve the business goals. The cloud adoption plan should also include a description of the risks and challenges associated with moving to the cloud. It is important to understand these risks and challenges so that they can be mitigated.

Finally, bringing it all together, consider what applications are prioritized and how you will transition them to the cloud and produce a master plan with stakeholder engagement, migration activities, and staff hiring and/or training. The plan should include a timeline for the migration (obviously!). This timeline should be realistic, and it should take into account the risks and challenges identified in the plan. You should also include a budget and consider the impact on existing resources that will need training, something that is often missed.

While listing out all the tasks can be lot of work (and tedious), the good news is the CAF provides ample guidance on how to prepare plans and provides tools to help you get started. The cloud adoption plan generator is a tool that can generate a task list or backlog within Azure DevOps.

Ready

This phase focuses on the steps to design a suitable cloud operating model and technical architecture to host the systems, applications, and workloads that will be part of the adoption project. Cloud architectures need to be optimized to leverage the right resources, cost structure, and capabilities for the needs of the specific organization and how it intends to operate the cloud.

The readiness checklist (a Word document) can be downloaded from the CAF website. It is a short, one-page guide with links to advice, tools, and best practices for designing and setting up your cloud.

Operating model

The cloud has redefined the way IT operates, but there is no one-size-fits-all model. Instead, the cloud demands an operating model that is tailored to the specific needs of the organization and capable of realizing the full potential of the cloud. The first step in designing an effective operating model is understanding the options available. There are two common business IT operating models:

- **Centralized**: In a centralized model, all IT resources and services are centrally managed and controlled. This model is best suited for organizations that prioritize the central control of architecture and technology decisions and likely strive for a more homogeneous technology landscape of systems and applications.

- **Decentralized**: A decentralized model typically splits IT resources and services among different business units or departments. This model is best suited for organizations that are motivated by innovation opportunities that come with greater business agility. Allowing departments to make their own technology choices removes any potential central bottleneck but may result in a more heterogeneous technology landscape.

Once you have a basic understanding of the options available, you need to define the requirements for your operating model. Common requirements include the following:

- **Location**: The first question to answer is where will the IT assets be located? Will there be mixed locations across a public cloud and private data center?

- **Governance**: How will resources be allocated and controlled? Who will have access to which resources? How will changes be made?

- **Security**: How will security be ensured? What are the threat vectors? How will data be protected?

- **Data migration**: How will data be moved to and from the cloud? What are the steps involved in migrating data?

- **Operations**: How will day-to-day operations be managed? What processes need to be in place?

- **Finance**: How will costs be tracked and payments made? What is the cost model?

- **Support**: Who will provide support for users? What about for systems?

The answers to these questions will help you design an operating model that fits the specific needs of your organization.

> **Take a step back**
>
> Please, do not skip the previous list of questions or just read through them. These are topics that you have to give deep thought and have great understanding of to be able to answer them. Stop. Pause. Think. Only after you have gone down the rabbit hole of each individual answer and understand the depth and breadth of the possible answers can you attempt to answer them.

Azure setup

To start using the Azure cloud, there are some basic things that need to be set up before you dive in and build or migrate anything. Unsurprisingly (perhaps?), the CAF includes a setup guide that covers starter topics such as Azure tenant and account setup. The guide also provides introductions to the tools available for managing resources, user access, costs, and billing. The Azure setup guide is available to read on the CAF website but is also presented as an interactive tool within the Azure portal Quickstart Center.

Not every scenario or organization is the same, so you may want to spend some time understanding the different options and tools before diving in. Familiarizing yourself with the basics can help in getting the initial setup right at the start and possibly avoid roadblocks and rework during the adoption phase. But don't let perfect be the enemy of good; it may be hard to change but it's not impossible – this is the cloud after all.

> **You can't keep everyone happy**
>
> Making changes to the cloud is the easy part. Keeping people happy is not so easy. All architects want to define a solid foundation on which to build their cloud. In real life, changing the foundations of a house is harder after the house is built. So too is the basic configuration setup of the cloud. Need to move your subscription to a new tenant without any disruption? Doubtful. Implementing naming and tagging standards after migrating all your workloads? Good luck with that.

The landing zone

Beyond the basic setup, you need to start thinking about how the cloud will be governed. Cloud security is an obvious concern for any organization adopting the cloud. But how can you ensure security is designed into the cloud by default and not bolted on at the end of the journey, especially when it will be much harder to implement?

Policy-driven governance is a design principle that can be used to manage risk and ensure compliance with corporate governance standards. Centralized policy management helps ensure that all cloud resources are compliant with corporate security and compliance policies. For example, defining security policies for network access to VMs or storage accounts early can ensure cloud resources do not have insecure settings that allow unauthorized access to data. Centralizing policies enables you to quickly respond to changes in the environment and enforce standardization across the organization (`https://learn.microsoft.com/en-us/azure/cloud-adoption-framework/_images/ready/alz-arch-cust-inline.png`):

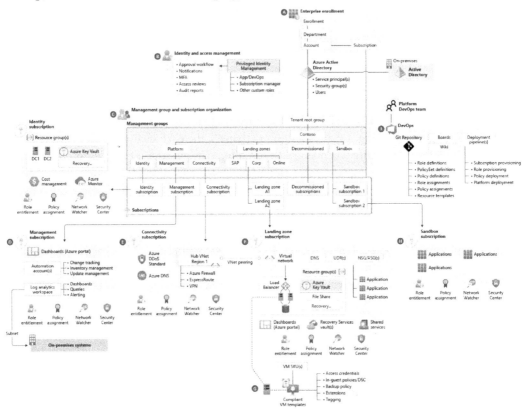

Figure 3.2 – An example Landing Zone design

An Azure landing zone defines the entire cloud environment for your organization, founded on centralized policy management (Azure Policy), core security services (Azure Defender), scalability, networking, and identity (Azure PIM). The CAF provides guidance on how to design a landing zone, including examples for different types and scales of organization.

Adopt

This phase is where the fun (and hard work) starts. You will begin using the cloud and aim to deliver on the business outcomes you promised everyone earlier on. The initial adoption project is likely to focus on either migration (if rehosting, or "modernization" if you plan to re-platform existing workloads) or innovation. Remember that it is advisable to prioritize a project that will deliver results quickly and give the organization time to learn. In subsequent iterations, it is possible that your project will increase in scope to achieve some migration and innovation in parallel. Cloud adoption is a long process, but architects can use the CAF guidance to plan and execute it successfully.

Migrate

The CAF reminds us that there can be a steep learning curve with cloud adoption. The first migration project should be seen as an opportunity to get familiar with the tools available – such as the aptly named Azure Migrate service. These tools can provide assistance with migration scenarios such as the "lift and shift" of virtual machines and databases, and in so doing help you choose the right size, high-availability configuration, backup strategy, and recovery options.

Innovate

The CAF encourages us to view the cloud not as an outsourced data center but as a platform for the creation of new digital inventions. In the cloud, you can leverage technology that was previously out of reach for most organizations. An extreme example of this is in quantum computing. Procuring, installing, and maintaining a quantum computer is not a practical proposition for most organizations today. But with a few clicks, you too can start developing code for Azure Quantum.

The sheer number of Azure services (200 and counting) can be somewhat dauting but can also be a lot of fun experimenting with. In many ways, on the cloud, the only barrier to innovation is our imagination.

Govern and Manage

Cloud adoption requires good management so that everything goes smoothly. We all endeavor to get things right at the start, getting Azure set up correctly, implementing the ideal landing zone architecture, automating everything, and achieving smooth, seamless migrations. But things don't always turn out exactly like we planned. The CAF defines the Govern and Manage phases as continuous and long lived, well after one might consider the adoption to be completed. Once you move to the cloud, you must ensure it stays compliant (either by design or policy) and that the cloud continues to deliver business value.

Most people shudder when they hear the word governance, unless they work in corporate IT where they eat governance for breakfast. Jokes aside, ensuring the cloud meets the business needs is everyone's business. Cloud governance is the practice of directing and controlling the use of cloud services. This governance encompasses all aspects of managing an organization's cloud resources, including provisioning resources, revising policies, monitoring costs, and decommissioning unused services.

Identifying and mitigating risks, ensuring the cloud continues to comply with organizational policies, and protecting the business's most valuable assets, all while enabling greater agility, are not to be treated lightly. Thankfully, the CAF provides tools to help you define a baseline for governance that can help you identify any gaps in your approach to governance. For organizations without any formal structure in place, the CAF provides guidance on methodologies and approaches that you can use to create a governance MVP.

The Manage phase of the CAF covers the breadth of traditional operations and forms the basis of a cloud operations function. The focus is on achieving the expected business outcomes, such as cost efficiency, availability, and performance. The CAF provides guides on Azure tools to help manage asset inventories, monitor SLOs, automate issue remediation, address threat protection, and much more. The CAF also provides a tool to create a management baseline that helps identify gaps in existing operational processes and practices.

The adoption journey

The CAF for Azure is a comprehensive resource that covers everything from how to get started with Azure to best practices for cloud adoption. Plus, it's always kept up to date with the latest learnings from Microsoft and experience from organizations that have used the framework. Best of all, it's completely free to use at the following page:

`https://docs.microsoft.com/en-us/azure/cloud-adoption-framework/`

At first glance, the CAF can be a little overwhelming – there is a lot of content. But didn't we say that a framework would simplify cloud adoption? Well, not exactly, as nothing worth doing is ever simple. But rest assured the CAF provides a sound structure for cloud adoption. How you choose to follow the framework and execute the phases defines your journey to cloud adoption. This is largely dependent on your business objectives and specific adoption scenario – migrate, modernize, or innovate. Regardless, there is nothing that mandates a waterfall journey. It is better – in fact, it is crucial – that you approach this with an agile mindset.

Remember to choose a path that promotes responding to change over following a carefully laid-out plan. Although a starting plan is critical, don't forget to be flexible and adjust as needed. Remember that people are more important than documentation, so have face-to-face conversations whenever possible. Ensure the documentation evolves with your adoption journey. Lastly, aim to deliver in short increments, to show value to the business frequently, rather than create radio silence and then a big reveal after several months or years.

To help you visualize what an agile adoption journey looks like, we will now walk through our recommended staged approach.

Form teams

Without a team, you won't be successful. Adoption requires people with various skillsets to achieve success. This stage typically involves building teams as required, so that everyone has a specific task to complete. Even though it may seem like a lot of work in the beginning, clearly identifying the roles and responsibilities (and then refining the organization RACI chart) of your current employees will ease the process significantly. We suggest four steps to this process:

1. Identify the need for teams.

2. Assess the skills and knowledge of existing personnel.

3. Define clear roles and responsibilities.

4. Build teams that have specific cloud adoption objectives.

Start with a team that can define the cloud strategy and high-level plans. This team could be *virtual*, meaning that people from different business teams can be assigned a role and responsibility that is not within their typical day job. This can be a great approach to getting buy-in from stakeholders early on by making them part of the process. Of course, it has the possible downside of delaying the very first step to adoption, as people try to balance commitments.

As the CAF adoption phases progress, form additional teams to get the work done. Throughout, keep in mind that changes to any organization can encounter resistance, often the subtle kind, but it can become pervasive if not handled correctly. Avoid creating silos, that is, teams that are disconnected from the rest of the organization. This limits the flow of information and can breed discontent in the teams that are keeping the lights on – that is, the team responsible for the current IT operations may feel left behind.

Instill a sense of ownership in teams by enabling them to make decisions and solve problems. Trying to control everything through top-down decision-making never works. At best, it can slow down your adoption plan as teams just wait for someone higher up to make a decision. At worst, the adoption never accomplishes the intended objectives – people generally find a way to stop something they don't believe in, or if they feel it is forced upon them. By empowering your teams, they are more likely to make delivery commitments and stick to them.

Don't forget that Microsoft has an extensive partner network that can provide experienced resources to help you at every stage of your journey. Augmenting existing staff and forming teams with expert partners can de-risk the adoption. But there are several potential disadvantages of outsourcing key aspects of your cloud adoption. The first, and most obvious, is the loss of control. When you outsource to a partner, you are trusting them with some aspect of your environment and operations. If something goes wrong, it can be difficult to get the plan back on track. Second, there is the risk of becoming overly dependent on the partner. What happens if they are no longer available or if their services become too expensive? Third, there is the question of knowledge transfer. Can you be sure that the partner will adequately transfer knowledge to your team? And finally, there is the question of culture. Will the partner fit in with your culture and be able to work effectively with your team?

To make CAF adoption successful, it is important to create teams that can further the cloud adoption journey. Identify the need for these teams, assess the skills and knowledge of existing personnel, define clear roles and responsibilities, and use partners judiciously. With the right mindset in place, you are well on your way to success.

> **Minimize dependencies**
>
> Ensure your teams can be represented as a list. If you need a matrix or anything more complex than a list with a team name and a description of the tasks they are responsible for, go back and simplify your team structure. And remember this: it is hard to manage dependencies between teams. Aim to create teams that can work independently with no dependencies on other teams.
>
> Collaboration is still important but getting work done in an independent way is key. Collaborate on best practices and ideas with customers and stakeholders – do not collaborate on managing dependencies between teams. Teams must be able to deploy independently and having dependencies – *any* dependencies – will prevent that. And dependencies multiply. So, the number of dependencies a team has must be exactly zero.
>
> One common pattern to achieve this is that teams work on and deploy independent services and the only communication between the teams is through an API layer. This is a technical example, but the same goes for the business requirements. Teams must work independently in order to have any meaningful agility and feedback/feedforward mechanism so they know what they do works and if not can course-correct.

Align your organization

You *should* know why having an adoption strategy is important – but what about the rest of your organization? The key guidance from the CAF is you must have organizational alignment. Like the wheels of a car, misalignment can have disastrous consequences when taking sharp turns or hitting bumps on the road. Clearly documenting your goals, expected outcomes, and KPIs helps to create the alignment needed. But in our experience, few people really read the documents.

Expect to present the strategy to many different stakeholders in a variety of settings – meetings, company presentations, one-to-ones, and so on. Also, expect to receive a lot of constructive... feedback. This should be incorporated into your strategy document, revised, and presented again... and again. As the adoption progresses and you learn more, it is only natural that things may change – in the business outcomes, the plan, or the landing zone. Any change that impacts the expectations of your stakeholders must be communicated to them to maintain alignment.

So, who is going to pay?

A topic that often comes up when aligning the business to the adoption is cost – who will pay for all of this? What can exacerbate this is your organization's history – is IT a cost center? Engaging with the CFO/finance team can be an opportunity to rewrite history and change the perception of IT. As we mentioned earlier, the cloud is a paradigm shift for IT. Large upfront investment is no longer the norm, but the pay-as-you-go consumption-based model is – from CapEx to OpEx. Learn how to speak with the finance folk and train them in the ways of the cloud, for a happy, more aligned future together. In our experience, nothing makes a CFO happier than IT folk being concerned with cost optimization.

Organizational alignment is the process of ensuring that all parts of an organization are working together toward a common goal. In the context of cloud adoption, this means ensuring that everyone in the organization understands and supports the cloud adoption plan. There are several ways to achieve organizational alignment. The most important is communication. Make sure that everyone in the organization has access to information about the cloud adoption plan, and that they understand why it is important. You can also use tools such as dashboards and scorecards to measure progress and ensure that everyone is on track. Finally, it is important to have clear goals and objectives for the cloud adoption plan so that everyone knows what they are working toward.

A **Cloud Center of Excellence** (**CCoE**) is a cross-functional team (project managers, architects, finance, InfoSec, and DevOps) that can further align the organization to achieve the business outcomes. The CCoE should aim to connect strategy with adoption plan execution and then with governance and the long-term operation of the cloud platform. The key responsibilities of the CCoE include the following:

- **Platform**: Ensuring that the cloud platform is operating effectively and efficiently, meets all business requirements, is well architected, and is continually updated to meet changing demands

- **Automation**: Furthering the goal to automate everything and maintaining standards for infrastructure as code and application deployment pipelines

- **Governance**: Ensuring that the cloud environment is secure and that all organizational policies are followed

- **Advocacy**: Promoting the use of cloud technologies within the organization

In addition to these core areas, the CCoE also needs to have a team of skilled personnel that can provide guidance and support throughout the cloud adoption process. This includes architects, developers, security professionals, operations engineers, and other technical experts. The CCoE should also be able to provide (or organize) training for employees on the cloud platform and provide ongoing support throughout the adoption process.

In summary, cloud adoption is a complex undertaking that requires careful planning and execution. Organizations must take steps to ensure that everyone in the organization understands and supports the cloud adoption plan and has access to the necessary resources to help them succeed. Establishing a CCoE is a valuable step in this process, as it ensures that different functions collaborate to deliver the cloud adoption.

Adopt the cloud

Adopting the cloud is, most likely, an iterative stage. We mentioned that delivering value in short increments of weeks is better than over months or years. This means adoption activities such as application migrations can and should be broken down into smaller deliverables. This does not necessarily mean that each deliverable is ready for real-world use but that it should be a demonstrable piece of working software – on the cloud.

As the iterations progress, you and the rest of the teams involved will gradually gain more knowledge and expertise. Coupled with the frequent feedback coming from demonstrations, effective teams will optimize their work. This means they will get better at migrating workloads or innovating, improve time and effort estimations, anticipate roadblocks, and solve problems faster.

However, to ensure the adoption achieves the desired results, effective governance is required. Expect a permanent team to guide the adoption toward its end state – basically, it is not a good idea to regularly replace team members during any significant project. The CAF recommends that you avoid the temptation to invent your own governance methodology. Instead, follow the guides and best practices within the framework. First off, define the rules to guide the adoption. Rules (or principals, policies, and standards) should be derived from risks.

We have mentioned strategy more than once and how clear objectives are essential. But understanding the business and technical risk is just as important. At a high level, making the risk explicit, that is, documented, owned, and prioritized, should provide a shared understanding of the tolerance for risk, how sensitive the data flowing through the systems is, and just how mission-critical the applications are. Decisions made on risk mitigations can be converted into rules that can then be applied by the teams responsible for delivering the cloud adoption. Having rules helps to empower teams to make decisions and move fast, without breaking too many things.

The governance team will be responsible for monitoring adherence to or violation of the defined rules. This should ensure the adoption is on track and provides an early warning when it is not. Five common governance disciplines help you continuously improve the adoption and your cloud over the long term:

- **Cost management**: Pay as you go is great, but without the right tools and regular monitoring, costs can spiral out of control
- **Security baseline**: Most (if not all) businesses must adhere to data protection regulations at a minimum, and governance must ensure compliance and drive continuous improvement

- **Resource consistency**: You must promote the standardization of resource configuration and processes, through change management controls

- **Identity baseline**: Roles and permission management must be applied consistently and reviewed for compliance with established rules and best practices

- **Deployment acceleration**: With standardization comes automation, eliminating manual changes, mandating infrastructure as code, full observability, and so on

With these five disciplines as a guide, the team must brainstorm procedures that fit the business, initially to govern the migration and over time the execution of the cloud adoption.

Summary

In this chapter, we introduced the CAF for Azure, without getting too technical or Azure specific. That is because the challenges are predominately cultural, meaning people and processes are hard to organize and harder still to change. The CAF provides a comprehensive set of adoption tools and resources to help you make your cloud journey successful. It offers, for example, readiness assessment tools, Azure setup guides, and planning templates. The framework helps you map your existing infrastructure and software to the cloud and enables you to identify any gaps between your current state and desired cloud target state. The CAF is the perfect starting point to embark on your adoption journey and make it successful. The next chapter, on migration, will focus more on the tools and technologies that will be used to move existing assets from an existing data center environment to the cloud.

4

Migrating Workloads

In *Chapter 3*, we described how to assess your digital estate during the planning phase and decide on the target state for each workload on the cloud. In this chapter, we will walk through the migration journey for workloads that will be rehosted or refactored (modernized). If your organization has not identified or does not require any migration, then you can probably skip this chapter for now, though the final section on Azure DevOps might be interesting.

In this chapter, we're going to cover the following main topics:

- Migration scenarios
- Migration tools
- Networking
- Virtual machines
- Containerization
- Azure DevOps

Understanding migration scenarios

Cloud migration is the process of moving applications, data, and the supporting IT infrastructure from on-premises (or an outsourced managed data center) to a cloud platform. Cloud migration provides organizations with greater flexibility, scalability, and, potentially, cost savings while allowing them to keep up with the ever-evolving cloud technology landscape. It also enables organizations to focus more on innovation than on IT procurement and management as cloud services are managed by the cloud provider.

In *Chapter 3*, we described the need to thoroughly assess your digital estate and determine the best migration path for each workload. Remember the five *Rs* of rationalization: *rehost, refactor, rearchitect, rebuild,* and *replace*. These five Rs are the most common options, though it is likely that when one pictures migration in their mind, it would be along the lines of a lift and shift (that is, rehosting). Lifting and shifting existing applications and data as is to the cloud can be a complex process, especially if the applications are legacy systems that may not be compatible with the cloud platform.

In order to successfully migrate existing applications to the cloud, you will need to do a thorough assessment of the application and its dependencies. However, you may not know there is a cloud compatibility issue until after the migration has begun. So, it is important to treat the migration as an iterative process. Build up experience of cloud services by doing – move beyond the theoretical and start using the services. You can create an Azure account with free $200 credit for 30 days and experiment with popular services such as **virtual machines** (**VMs**). This can better inform the migration plan and help you identify workarounds or alternative solutions for issues when they arise. Refactoring applications (modifying the source code) or rearchitecting systems can result in major roadblocks to migration. Most workloads, though, comprise many components, including databases, file stores, and firewalls. So, a lift and shift could cover a large part of your workload, while some components (hopefully just a few) may require refactoring or complete replacement.

> **Refactoring**
>
> Folks get quite attached to the work they've done. It is sometimes difficult to discuss major refactoring because of this. In consulting with some of the largest enterprise companies in the world, this often comes up. Departments grow attached to their services; individuals get attached to their work and don't easily give up control or want to consider an alternative approach.
>
> When looking at refactoring, in an example where you are trying to refactor a monolith into a cloud-native serverless architecture, it might surprise you that you can achieve the same service function with a lot less code – all the boilerplate, all the additional services you needed to develop to get this service working now, might be included in the cloud services. For example, a large application that used to convert data into PDF documents might now be replaced by an API and a few very tiny functions and perform the same functionality, while ordering, downloading, and displaying the document is relegated to the calling service. Suddenly, you are going from a service that needed large teams to create to a service that a few people can maintain indefinitely. And you've gotten more resiliency than you might have had before.

Application latency is an important factor to consider when migrating workloads to the cloud. It is the amount of time it takes for data, instructions, and results to travel between the cloud application and its user – that is, a human end user or another component of your system. If cloud applications experience high latency, it can lead to slow performance or even outages. In order to ensure a smooth cloud migration process, organizations must take into account potential latency issues before and during the migration. Latency is also affected by network traffic, so it's important to plan out how applications will be distributed across cloud regions and availability zones. This will help reduce any unnecessary latency caused by having too much network traffic in a particular area. Additionally, cloud providers often offer services such as **content delivery networks** (**CDNs**) to optimize performance by caching data closer to end users, reducing the overall latency (`https://blogs.partner.microsoft.com/mpn-canada/wp-content/uploads/sites/81/2017/06/Microsoft-Azure-Cloud-Models.png`):

Cloud Models

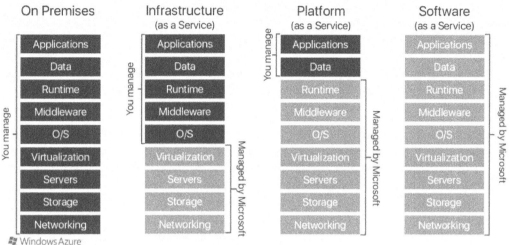

Figure 4.1 – Cloud models

When rehosting existing workloads (including both applications and data) on the cloud, a common scenario is to use **infrastructure as a service (IaaS)** as it closely resembles existing IT infrastructure but on the cloud. IaaS is a cloud computing model in which the cloud provides virtualized computing resources such as servers, storage, and networking over the internet. It enables organizations to deploy cloud-based applications and IT infrastructure quickly, easily, and cost effectively, without having to purchase physical hardware or manage its own data centers. IaaS for cloud migration includes components and concepts such as VMs, file-based data storage disks, and network interface cards. VMs can be used to quickly launch cloud-based applications with the same functionality as their on-premises counterparts. Operating systems provide the flexibility to customize cloud instances based on specific needs. File stores provide access to large datasets securely from any location worldwide and firewalls protect the cloud infrastructure from malicious activities. Additionally, cloud automation tools such as autoscaling can ensure your applications always stay optimally allocated to match changing workloads. IaaS also offers more robustness compared to traditional on-premises setups due to its ability to replicate data across multiple availability zones, making it highly resilient against outages or system failures. Furthermore, IaaS solutions are usually cost effective since organizations only pay for what they use without having huge upfront costs in setting up their own data center infrastructure. When rehosting, consider also that a lot of services, websites, and data that gets provisioned on-premises and is then kept forever might actually be candidates for deletion (or we can call it archiving) rather than actual rehosting. This is a great time to find the owners of each and find out whether this is really needed in the organization going forward. IaaS cloud solutions, while cost effective and offering many advantages, also have a few drawbacks that must be taken into consideration. One of the primary drawbacks of IaaS cloud migration is the lack of visibility. Since cloud resources are virtualized, it can be difficult to monitor the performance or usage metrics in order to identify any potential issues – unless you address these and use a service such as Azure Monitor from the start to get deep insights. Additionally, cloud resources are typically provisioned on-demand and may not have the same level of reliability as what you are used to with physical hardware. This could lead to higher or lower service levels for applications hosted in the cloud compared to their counterparts running on traditional infrastructure. Furthermore, cloud vendors often charge for compute and storage resources separately, making it more difficult to accurately predict future costs and manage budgetary constraints. All we are saying here is that there are differences between virtual and hardware-based models and to pay attention to the differences and account for them.

Another concern with cloud migration is security. While the cloud offers advanced security measures, such as encryption, identity and access management, and **intrusion detection systems** (**IDSs**), they still rely on customers to configure their cloud environment correctly and maintain a secure environment through regular patching and software updates. Furthermore, migrating data from an on-premises environment to a cloud platform requires additional steps, such as data masking or encryption, in order for sensitive information to remain protected against unauthorized access or other malicious activities. Finally, due to its virtual nature, IaaS cloud environments are sometimes less capable (in terms of compute power or memory efficiency) than dedicated physical servers when it comes to certain specialized workloads, such as gaming or **high-performance computing** (**HPC**) unless you use specific services that address these issues.

Another scenario for cloud migration is to leverage more **platform as a service** (**PaaS**), a layer above IaaS. PaaS enables IT infrastructure on the cloud but without having to manage all aspects. For example, using application hosting services, database management systems, and analytics and monitoring tools is ideal for organizations looking to quickly migrate applications without having to worry about the complexity of setting up or managing the services using VMs.

PaaS cloud solutions typically offer the same scalability compared to IaaS deployments due to the agility of cloud servers that can be rapidly deployed and deactivated as needed. Additionally, PaaS often provide high-availability features, such as load balancing and automatic scaling, which helps to ensure applications are always running optimally, but with less configuration management overhead. PaaS also offers a number of security benefits, such as automated patching and updating, encryption of data at rest and in transit, identity management tools, **access control lists** (**ACLs**), and IDSs. These components help protect cloud environments from threats such as data breaches or malicious attacks. Additionally, some PaaS solutions also provide advanced compliance capabilities that allow customers to meet industry or government regulations when deploying cloud-based applications.

Finally, PaaS solutions are particularly suitable for organizations looking to modernize legacy applications by transitioning them into cloud-native architectures without requiring significant effort or investments in rearchitecting the code base. With the right migration strategy and approach, it is possible for organizations to quickly move existing applications onto the cloud while opening the door to additional functionality on offer, such as machine learning algorithms or **artificial intelligence** (**AI**) models that can improve the user experience when interacting with the application. Next, let's dive into the tools you can use for cloud migration.

Examining migration tools

Microsoft Azure provides numerous tools and services for organizations looking to migrate their workloads into its cloud platform. On top of this, there are numerous Microsoft partners (**independent software vendors**, or **ISVs**) offering tools and services to facilitate migration as a whole or targeting specific technologies or applications. The Azure Migrate hub is a unified experience for organizations that brings together all these tools, enabling you to quickly, securely, and cost-effectively move applications and data to the cloud. With Azure Migrate, organizations can discover, assess, migrate, and manage their workloads in one place (find the original image at `https://learn.microsoft.com/en-us/azure/cloud-adoption-framework/migrate/azure-best-practices/media/migrate-best-practices-costs/assess.png`):

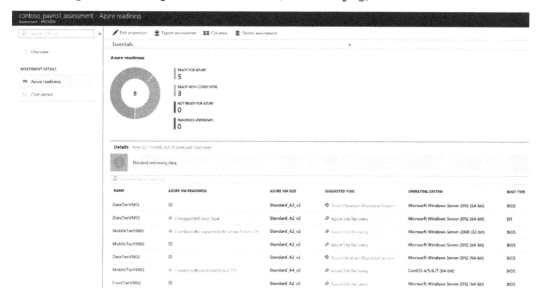

Figure 4.2 – Azure Migrate screenshot

Azure Migrate simplifies the cloud adoption process and helps customers save time by automating various steps, such as discovering applications running on-premises and identifying dependencies between application components. It also provides detailed insights into cloud readiness levels for each component before migration begins, allowing you to make informed decisions about what needs to be done before beginning the process. In addition to helping assess potential migration projects, Azure Migrate also facilitates cloud migration through its integrated interface with other Azure services. For example, it can connect with Azure Database Migration Service, which enables rapid and reliable database migrations from multiple sources to cloud databases hosted in Microsoft Azure. Additionally, users have access to step-by-step workflows that guide them through common cloud scenarios, such as moving servers to VMs or migrating web applications using containers.

Azure Migrate can be accessed directly from the Azure portal. To get started, you will select the migration goal: servers, databases, and web application; databases only; VDI; web applications only; and so on. The process of migrating servers begins with creating a new migration project. This project will hold all the metadata about your on-premises environment generated by the discovery and assessment tool you select.

The built-in **Server Assessment** tool from Azure can be configured to connect into your environment. If you are running System Center Operations Manager or are already virtualizing servers with the Hyper-V or VMware platform, the Azure Migrate assessment tool can communicate with these and collect metadata on servers, such as naming, operating system details, and sizing (CPU and memory). If not, you can install an agent on your servers that communicates with the assessment tool. The output of the assessment tool can be invaluable, by quickly visualizing the scope of migrations, including dependencies, and could uncover things you may not have known about your environment.

The web application assessment tool can generate metadata such as the web technologies and frameworks used and determine whether the application could be hosted using Azure App Service. If the web application is available publicly, you can run the assessment directly using the public URL. If not, you can download a tool to run directly against your .NET, PHP, or Java application. The following image can be found here – `https://learn.microsoft.com/en-us/azure/architecture/solution-ideas/articles/azure-vmware-solution-foundation-capacity`:

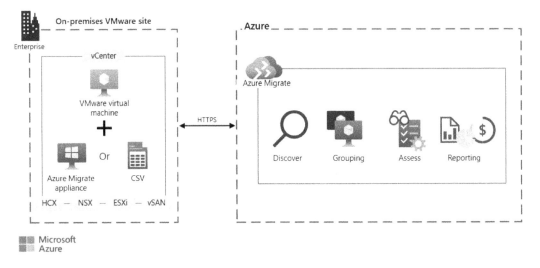

Figure 4.3 – On-premises migration concept

Once you have completed the discovery and assessment, next you select the migration tool. The built-in migration tool (that is, provided by Azure, but partners also provide tools) automates the process of replicating on-premises servers to cloud VMs and guides you through testing and executing the final migration. The complexity of cutting over to your new cloud servers will be very specific to your environment and system architecture, but the migration tools perform a lot of the heavy lifting, allowing you to plan this out and test with your users relatively quickly.

Another noteworthy tool is **Data Box**, a cloud data migration service offered by Microsoft Azure. It enables organizations to quickly and securely move terabytes of data from on-premises storage systems into the cloud. This is done via a Data Box device that you can order through the Azure portal. Once received, it can be configured locally and begin the transfer of data from your on-premises systems to the device. Once the local transfer is completed, you can return the device and the data will be uploaded to Azure. This is particularly useful in scenarios where there is limited or slow network connectivity or where the cost of transfer of large datasets would be expensive. Another benefit is security, particularly of sensitive data transferred over the public internet, which is too high a risk.

Overall, Azure Migrate streamlines the cloud migration process and provides everything needed in order to successfully complete IT migrations while reducing the risk associated with deploying critical business applications into a new environment. Next, we will review networking within the context of the cloud.

Understanding networking

Cloud networking is an essential part of cloud migration and application modernization. It provides a way for servers and platform services to connect and share information and potentially communicate with on-premises systems, enabling organizations to access cloud resources such as applications and data storage from anywhere in the world. Microsoft Azure provides a comprehensive set of cloud networking solutions to meet the needs of most IT migration projects. These solutions include **virtual networks** (**VNets**), secure site-to-site connections, load balancers, and DNS servers.

An Azure VNet establishes a dedicated, isolated, and highly secure environment within the Microsoft Azure cloud, allowing you to control how data flows between different cloud resources. Networks are subdivided into smaller networks (subnets), providing administrators with the capability to isolate network traffic using IP address routing prefixes or masks. This hierarchical organization and segmentation help to logically manage the network in order to increase security and separate applications and workloads (find the original image here – `https://learn.microsoft.com/en-us/azure/architecture/networking/spoke-to-spoke-networking`):

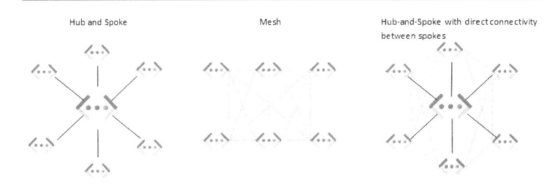

Figure 4.4 – VNet design patterns

VNets can be assigned access policies to control incoming and outgoing network traffic, similar to traditional network firewall appliances. These utilize application layer filtering rules, which allow organizations to control inbound and outbound traffic within their VNets. Internet protocol and HTTP header inspection rules can be easily defined to block certain traffic and allow others. This allows administrators to enforce business policies at the network layer while also locking down public access to cloud resources. For example, permitting access only to a web application's HTTPS port or blocking an application from transferring data to a server on the internet could indicate malicious activity.

Azure **network security groups** (**NSGs**) are a solution that act as virtual firewalls to control inbound and outbound network traffic. They provide an additional layer of security to your Azure resources by allowing you to define rules to filter traffic based on the source, destination, and protocol. Azure Firewall is another cloud-based network security service that provides advanced threat protection for your applications and network assets by filtering traffic at the application and network levels. It allows you to create application and network rules to block unwanted traffic and provides outbound connectivity to external resources. Azure App Gateway is a web traffic load balancer that manages traffic to your web applications. It provides features such as SSL offloading, cookie-based session affinity, and URL-based routing. These products can be used together in different scenarios to ensure optimal protection and performance for your applications and services. For example, NSGs can be used to easily implement IP address and port access restrictions, Azure Firewall can provide more powerful traffic restrictions, and App Gateway can be used to route traffic and improve application performance.

A **virtual private network** (**VPN**) can be defined to enable connection between on-premises networks and cloud-based resources by establishing secure site-to-site connections via IPsec VPN gateways or ExpressRoute circuits. A VPN allows network administrators to securely extend network services from one location to another over an unsecure network, such as the internet. With a site-to-site VPN, data is encrypted and travels through a secure virtual tunnel between two endpoints, providing more efficient setup and management over establishing individual physical point-to-point connections between two network sites.

> **But I thought VPNs were for privacy?**
>
> While here we are talking about establishing trust between two sites and communicating across the public internet, most people use a VPN to protect their online identity or to access content not available in their country. One thing that people often don't understand about VPNs is that while they can help protect your online privacy and security, they're not a foolproof solution. While a VPN can encrypt your internet traffic and mask your IP address, there are still ways for third parties to potentially track your online activities – such as cookies, unscrupulous browser add-ons, and, if you are unfortunate to have it, malware. It is important to protect your device, carefully select and install software from reputable sources, and choose a VPN provider that doesn't log user data or engage in shady practices with your personal information.

Azure also supports the definition of load balancers to distribute traffic across multiple cloud resource instances and regions, allowing organizations to scale their applications horizontally. It works by routing network requests based on predetermined rules to applications that can best respond to the network request. This eliminates issues caused by application bottlenecks and ensures that users have uninterrupted access to the applications they intend to use. An application load balancer can reduce network latency and increase scalability, enabling smoother operation and faster response times over a longer period of time, making it a valuable tool for cloud-hosted applications and services. Azure load balancing supports Layer 4 (transport) and Layer 7 (application) protocols, allowing you to balance traffic across multiple VMs or instances based on various criteria, such as round robin, IP affinity, or least connections. Load balancing can be implemented at different scopes, including regional and global. Regional load balancing distributes traffic within a specific region, while global load balancing can distribute traffic across different regions. The choice between regional and global load balancing depends on factors such as the location of your users and the level of redundancy required for your applications. When implementing global load balancing, you should consider factors such as DNS resolution, health monitoring, and failover mechanisms. Azure also provides additional features such as SSL offloading, TCP and HTTP health probes, and connection draining.

Check out this link for more information: `https://learn.microsoft.com/en-us/azure/architecture/guide/technology-choices/load-balancing-overview`.

As with all Azure services, there is extensive monitoring out of the box. Network monitoring is an essential activity for IT teams to ensure availability, uptime, and performance. Typical network monitoring use cases can range from basic network health checks as a preventative measure to more complex network troubleshooting activities, such as request load and failures. Network monitoring across cloud resources typically involves reviewing network interfaces, routing rules, and access policies. Additionally, detailed traffic log files may be monitored to review active sessions, filter errors and warnings, and detect any malicious or suspicious traffic on the network. Anyone with experience of network administration knows how complex this activity can be even for moderately sized systems.

Microsoft Defender for Cloud is a security solution that uses industry-standard best practices and a variety of advanced data analytics to detect, investigate, and protect against malicious threats and keep you safe from attempted network intrusions. With just a few clicks in the Azure portal, you can get complete visibility into your network in near real time, while Defender's security monitoring service can help identify suspicious behavior or anomalous traffic that may signal an attack.

Finally, Azure networking is designed with resiliency in mind; it is built with a redundant architecture that allows it to remain operational even if one of the underlying data center components fails temporarily. Now, let us discuss the most important resources in the cloud, compute power or VMs.

Understanding VMs

VMs are the bread and butter of any cloud migration, allowing physical machines to be displaced by, well, virtual ones! VMs are not a new technology by any measure, but the scale at which they operate on the cloud is incredible. With just a few clicks, you can create a machine anywhere in the world and log in after a few minutes. Need multi-core CPU, 128 GB of RAM, a 1-TB disk, and a gigabit network interface? No problem, use the drop-down menu. Need a specific (modern) operating system version? No problem, install it using a predefined operating system image.

VMs achieve this magic by using a hardware abstraction layer. VMs facilitate the transfer of applications to the cloud with minimal restrictions on the machine specifications or performance penalty due to its hardware abstraction. If your application runs today on an Intel/AMD x86 64-bit machine, it can be lifted to an Azure VM and not notice any difference… probably. Yes, there are a few gotchas, particularly if your application developers integrated somewhat niche physical hardware modules or use obscure, low-level APIs – security modules and Linux device I/O come to mind. Hopefully, some application code refactoring can resolve any issues you encounter. It would certainly be worth never having to worry about hardware upgrades again.

> ### A machine inside a machine
>
> One thing that people often don't understand about VMs is that they are essentially emulating an entire computer system within another computer system. This means that the performance of the VM can be impacted by the resources available on the host machine, and that certain hardware features may not be available within the VM environment. Additionally, it's important to remember that any data stored within a VM is still vulnerable to security threats, so it's important to take appropriate measures to protect your virtualized systems just as you would with physical ones.

By using VMs deployed in cloud architecture, companies can save resources while gaining agility compared to on-premises solutions. For example, clusters of VMs may be created for test/dev and production workloads, allowing various configurations to be deployed. This flexibility also allows architects to develop resilience strategies with multiple redundant servers for maximum uptime. VMs can undeniably help server administrators to automate server management processes, allowing them to spend more time on tasks with higher value. Unlike a physical server that is tied to a particular hardware, VMs can be configured and moved without any impact on operations or applications running on the server. With virtualization technology, server administrators have control over server resources and can change this as and when it is required. Ultimately, deploying virtualized servers enables organizations to cost-efficiently increase agility while continuing to get the performance they need from server workloads.

Azure VMs are cloud-based computing resources that provide users with access to virtualized server environments, enabling them to deploy and maintain applications without needing to manage physical hardware. VMs come in a variety of sizes and configurations, allowing users to customize their environments according to their application's demands. With VMs, cloud administrators can easily provision, configure, scale, and manage cloud-based services while maintaining the same performance they would expect from physical hardware.

Using cloud virtualization technology, cloud administrators have full control over server resources such as CPU cores, memory, disk storage, as well as network bandwidth. Administrators can create clusters of VMs to configure different configuration requirements, such as dedicated test/dev or production workloads for maximum flexibility. This means that cloud administrators can define performance parameters for their applications in order to meet specific **service-level agreements (SLAs)**.

When deploying cloud VMs on Microsoft Azure cloud platforms, users gain access to advanced cloud security tools, such as Microsoft Defender for Cloud. Defender uses industry-standard best practices along with automated data analytics and intelligence capabilities that help detect malicious threats before they can cause damage or disrupt operations. Additionally, since VMs are based on cloud architecture, they benefit from having built-in resiliency (if you plan for it correctly), which allows them to remain operational even when underlying data center components fail temporarily.

In conclusion, VMs on the Microsoft Azure cloud platform offer an array of advantages compared to traditional on-premises solutions. Not only do they free users from the burden of managing physical hardware but they also provide high resilience and scalability at a fraction of the cost associated with on-premises solutions. Next, let's consider another option for managing compute resources.

Examining containerization

Application containerization is revolutionizing the way we think about and manage distributed computing workloads. Containers are lightweight, portable units of software that can run anywhere and contain all the necessary dependencies to run an application. By abstracting applications away from their underlying infrastructure, containerized applications are more agile and easier to move from one environment to another with minimal disruption. This allows organizations to quickly provision new compute resources and redeploy existing ones in order to respond faster to changing demands or market conditions. Sounds a lot like VMs. With VMs, you still have to manage the operating system: configuration, patching, storage permissions, network interfaces, application installation, and so on. A major challenge of VM administration is standardization.

Pets versus cattle, who wins?

Within software architecture, the standardization challenge is, in my opinion, best described using the analogy of pets versus cattle. Pets are unique and special; they have a name, require special food, have frequent trips to the local grooming service, and receive lots of love and care from their owner. The loss of a pet can be quite traumatic and such a loss can never be replaced. Application server "pets" require more manual work to maintain and operate. They typically don't resemble any other application and so extra care is required when rolling out operating system patches or configuration updates. This often occurs because the developers were given free reign over the server they run on. Losing an application server pet can be quite traumatic to an organization. In most cases, the original developers are the only people with the specific skills and knowledge to recreate the server. Cattle, on the other hand, are less individualized; they perform a number of functions and can be easily replaced. Application server cattle are grown from clear design principles that standardize the operating environment and promote the automation of configuration management and deployment. Such server cattle can be spun up, scaled out, and shut down, as well as crash and fail over, with minimal to no fuss.

One way to achieve configuration standardization is by containerizing our applications. A container (analogous to a shipping container) is a standard packaging of application code, configuration, and all its dependencies – everything required to run the application. Containers are executed by a container runtime engine, another layer above the hardware abstraction layer that hides or abstracts many aspects of the operating system. Software can be quickly distributed and deployed without the need for the manual configuration or provisioning of servers. This allows a containerized application to run anywhere regardless of the underlying infrastructure – a developer's laptop, on-premises server, VM, or cloud container service. This makes it ideal for cloud migration and application modernization projects. Developers can then focus on building applications and easily package their software in containers, safe in the knowledge that their applications will run in any environment.

Never again will we hear, "*Well, it worked on my machine.*"

IT admins, on the other hand, can enjoy a more consistent standardized environment and care less about the applications that run on them. However, taking responsibility for the installation, configuration, and operation of a container environment can come with a steep learning curve. Containerization adds an additional layer of complexity to the deployment and management of applications, which can be difficult for teams without experience in this area. Containerized applications require careful network configuration to ensure that different containers can communicate with each other securely and efficiently. In most cases, containers share a host operating system with other containers, which could be beyond the comfort zone of traditional IT admins to manage and could lead to security risks and potential vulnerabilities if they are not properly configured or managed.

Docker, an open source containerization platform, has become the de facto standard. Microsoft Azure offers native support for deploying Docker containers on its cloud platform, making it easier than ever to migrate existing workloads to the cloud with minimal effort and less overhead for IT admins – if the workloads are already containerized; otherwise, some more work is required. Azure provides a number of container PaaS services that support running containerized applications in different ways (with somewhat overlapping features) depending on the specific needs of an organization or workload. The following are examples:

- **Azure Container Apps**: Allows you to build and deploy modern applications and microservices without the overhead of managing complex infrastructure

- **Web App for Containers**: Specifically designed for modern web applications and APIs with seamless support for CI/CD

- **Azure Container Instances**: Offers hypervisor isolation when security requires no sharing of the host operating system kernel

- **Azure Service Fabric**: Designed for complex distributed systems where reliability, scalability, latency, and ease of management are critical to successful business operation

What is container orchestration?

Container orchestration is the process of automating the deployment, management, scaling, and networking of containers. Container orchestration platforms provide a way to manage large numbers of containers across multiple nodes or servers in a cluster. This includes tasks such as deploying and running (or scheduling) containers on specific cluster nodes, monitoring container health, scaling up or down based on demand, and managing container networking so that different containers can communicate with each other seamlessly. Container orchestration makes it easier to deploy and manage containerized applications at scale and ensures that they are highly available and resilient to failures.

> **Kuber what now?**
>
> If you work in IT, I will bet money you have heard about **Kubernetes** (also known as **K8s**), the open source container orchestration platform that automates many of the manual processes involved in deploying, managing, and scaling containerized applications. It was originally developed by Google and is now maintained by the Cloud Native Computing Foundation. Kubernetes allows you to easily manage and scale containerized applications across a cluster of nodes, providing features such as load balancing, automatic failover, self-healing, and rolling updates.

Azure Kubernetes Service (**AKS**) is a cloud-based container runtime and orchestration service. It provides an optimized configuration of popular open source tools and technologies for deploying and managing containerized applications and microservices. AKS allows for the simplified and rapid creation, deployment, scaling, and management of containers on cloud infrastructure. AKS uses Kubernetes as its orchestration engine to provide a secure, highly available platform that can be used to deploy cloud applications with minimal effort. It supports both public cloud-based environments such as Azure cloud services and private cloud-based environments such as Microsoft's own Windows Server 2016 containers. With AKS, you can easily create complex application architectures using preconfigured templates or custom code. You can also manage multiple cloud environments with a single console across all your deployments – enabling unified administration and access to all workloads running in the cloud. Using AKS, you can achieve high scalability for your cloud applications by dynamically provisioning new compute resources when demand increases and deprovisioning them when demand decreases. This helps you save costs by ensuring that resources are only allocated when needed while continuing to provide a consistent application experience for users. Additionally, AKS enables the more efficient use of resources through automated allocation across different compute types according to specific workloads.

Exploring container registries

A container registry is a centralized inventory of container images that can be used to deploy applications in containerized environments. It's a key component of the container ecosystem, as it provides a secure and scalable way to store and distribute container images. Container registries allow developers to store these images in a centralized location where they can be easily accessed by other team members or deployed to cloud environments. Container registries typically provide features such as the following:

- **Image storage**: Container registries provide a place to store container images securely
- **Versioning**: Container registries enable the versioning of images so you can keep track of changes over time
- **Access control**: Container registries allow you to control who has access to your images
- **Image scanning**: Some container registries offer security scanning tools that scan your images for vulnerabilities or compliance issues
- **Replication**: You can replicate your images across multiple geographic regions for faster deployment times and increased resilience

Some popular container registry services include Docker Hub, Google Container Registry, Amazon **Elastic Container Registry** (**ECR**), and Microsoft **Azure Container Registry** (**ACR**).

Docker Hub is a cloud-based container registry provided by Docker that allows developers to store, share, and manage Docker images. It's the world's largest library and community for container images with over 7 million images available. Docker Hub provides a centralized location for storing and sharing Docker images with other members of your team or the broader community. You can search for new container images in Docker Hub's vast library of pre-built images that cover a wide range of programming languages, frameworks, and tools. Docker Hub also integrates with other Docker tools, such as Docker Desktop along with Kubernetes and the Azure container services generally.

Similarly, ACR is a cloud-based container registry service provided by Microsoft Azure that enables developers to store, manage, and deploy Docker container images. It's a fully managed service that provides features such as private repositories, secure image signing and scanning, and integration with other Azure container services. ACR allows you to securely share your container images with other members of your team or organization using **role-based access control** (**RBAC**). ACR integrates seamlessly with Azure DevOps and other Azure container services, such as AKS and App Service, making it easy to incorporate containerization into your existing development and release workflows.

A container registry is an essential tool for developers working on cloud-native applications using containers as it simplifies the process of storing, managing, securing, and deploying containerized applications in the cloud. Using Azure container services helps to eliminate infrastructure complexity at scale while providing an easy way to run distributed applications in the cloud, as well as minimizing the management of underlying compute resources. Now, let us switch gears and dive into tools to help manage work tasks and source code and further automate cloud-based delivery.

Looking at Azure DevOps

When looking to migrate workloads into the cloud, it also makes sense to help yourself by automating development, integration, and deployment pipelines. While Azure DevOps services are cloud agnostic, tie-ins with Azure make Azure DevOps a great way to achieve some of that sweet automation. **Azure DevOps Services** (**ADO**) from Microsoft is a cloud-based SaaS solution for application life cycle management, from planning and development to QA and deployment. With ADO, you can easily develop applications in the cloud and deploy them on any platform – not just the cloud. ADO helps developers quickly develop, ship, and maintain applications with a wide range of tools for software development teams. It enables easy collaboration between developers working in different locations or across multiple teams. Teams can create projects that are automatically provisioned with all necessary components, such as the source code repository, build pipelines, and test plans and cloud resources.

Why look into ADO specifically? One reason is that it is a tool that can complement and assist with your migration (and other) goals.

Using ADO **continuous integration/continuous delivery** (**CI/CD**) features, you can quickly deploy applications into production with automated builds and releases that ensure the highest quality results within the shortest possible timeline, as well as cost reduction. Teams can collaborate on projects using a suite of integrated tools that complement an organization's existing practices and workflows. ADO comprises the following tools:

- **Azure Boards** helps you manage agile projects such as software development, cloud adoption, migration, and infrastructure automation projects. It provides a centralized place to track the work that needs to be done, assign tasks to team members, and track the progress of the project.

- **Azure Repos** provides source code repositories built on Git, a code management and version control system that helps you track and review changes to your source code.

- **Azure Pipelines** features CI/CD capabilities that enable fast deployment into production with automated builds and releases. It can help you speed up the process of getting your code from development to production.

- **Azure Build Agents** are processes designed to do the work of building and packaging software. You can choose Microsoft managed agents or connect your own VMs hosted in Azure. Agents can be managed directly from the ADO portal, enabling team agility by reducing dependency on IT teams for life cycle changes, build framework upgrades, and so on.

- **Azure Artifacts** enables developers to share their code packages and builds easily with their team, organization, or the wider public. Developers can manage any code dependencies from a centralized location with integrations with package registries such as NuGet.org or npmjs.com.

- **Azure Test Plans** provides an array of robust tools that every team member can use to collaborate while the development process is underway. The intuitive test management application comes with all the features necessary for managing automated tests, planning manual testing cycles, and gathering feedback from user acceptance testing. It is fully integrated with Azure Boards and Pipelines, enabling clear traceability from test to build to user requirements.

ADO's integrated marketplace is a great addition to the service, providing users with an ever-growing selection of powerful extensions and tools (2,040 extensions as of December 2022). These can be used to enhance workflows, extend the capabilities of pipelines, and enable integration with other tools and services.

Staying agile with Azure DevOps

"*Individuals and interactions over processes and tools*" is one of the four values outlined in the *Agile Manifesto*, which serves as a guiding principle for agile methodologies. This value emphasizes the importance of prioritizing people and communication over rigid processes and tools. It recognizes that while processes and tools are important, they should not be viewed as more important than the individuals involved or the interactions between them. In an agile context, this means that teams should focus on building strong relationships between team members, stakeholders, and customers. This may involve creating collaborative work environments where team members can openly communicate their ideas, concerns, and feedback. It also means that teams should be flexible in their approach to project management. While following established processes is important for consistency and quality control, it's equally important to recognize when a process or tool isn't working or needs to be adapted based on changing circumstances.

Scrum and Kanban are two popular approaches for managing agile projects. Both aim to help teams work more efficiently and effectively, but they have different approaches. Scrum is a framework that focuses on delivering working software in short iterations (typically two weeks in duration) called sprints. The team works together to plan out the work for each sprint, then holds daily sync meetings (aka a stand-up or team huddle) to stay aligned and address any issues. At the end of each sprint, the team reviews their progress and identifies areas for improvement. Kanban is a visual system for managing workflow. Instead of dividing work into sprints, it emphasizes CD by limiting the amount of work in progress at any given time. Teams use a Kanban board to track tasks as they move through different stages of completion, from "to do" to "in progress" to "done." Both Scrum and Kanban have their strengths and weaknesses depending on the project's context. Scrum can be great for complex projects that require frequent check-ins with stakeholders or that involve many moving parts. Meanwhile, Kanban may be better suited for simpler projects with fewer dependencies where CD is important, but it can successfully be used for complex projects as well.

Regardless of the methodologies you are using (or plan to use), Azure DevOps has you covered. Out of the box, there is support for agile processes, meaning the boards are ready to use with work item/task templates, ample data attributes (fields), status options, and agile workflows.

> **Flexibility**
>
> While flexibility is great, even highly desirable, take caution when customizing Azure DevOps to match how you manage agile projects. If you have many task statuses beyond to-do/doing/done, complex workflows, gates (mandatory checks and balances), and so on, it might be a sign of agile anti-patterns and time to reassess how you do things. Stick to the out-of-the-box templates if you can – your team will thank you, though management might not!

If you have used a tool to help manage your agile project, there is likely an easy way to migrate historical tasks and work-in-progress tasks to Azure DevOps. At the most basic level, you can import tasks stored in a CSV file. So, if your tool can export tasks to CSV, with some minor data transformation to match the import column spec, you can import them. However, unless you plan a "big-bang" migration and a clean cutover from your existing tool, it will be hard to keep multiple tools in sync using CSV files.

Jira is a popular project management tool developed by Atlassian. It is designed to help teams manage their work by providing a centralized platform for tracking tasks, bugs, and issues related to software development and other projects. Jira supports agile methodologies such as Scrum and Kanban with visual boards that allow teams to manage their work in progress. The ADO marketplace offers Jira connectors that can be used to keep the tools in sync. New Jira tickets and changes can appear within ADO in near real time. This offers a pragmatic approach to migration, allowing teams to become familiar with ADO and transition over to regular use when it suits them.

The Azure Cloud Adoption Framework provides ADO project templates that you can use to create migration plans. The templates come with sample content that includes source code, backlog work items, predefined planning iterations, service endpoints, and build and release definitions. ADO is the perfect toolset to help your team accelerate cloud migration processes and modernize applications quickly. For more information, visit the Cloud Adoption Framework *Tools and templates* page: `https://learn.microsoft.com/en-us/azure/cloud-adoption-framework/resources/tools-templates`.

Migrating your source code

There are many other version control systems out there, but the most common, in our experience, are **Concurrent Versions System** (**CVS**), Apache **Subversion** (**SVN**), and, dare we say, Microsoft **Visual SourceSafe** (**VSS**) – yes, there are legacy systems still depending on a discontinued version control system! If your organization has source code locked away in one of these systems, it is high time you considered migrating to Azure Repos. We encourage you to explore the Azure DevOps marketplace and experiment with the migration tools available. However, if you don't find anything that suits you, find out whether the change history is vitally important. Chances are it holds more sentimental value than actual usefulness, so you can always import the latest source code version and enjoy a fresh start with Git.

Azure Repos is built on Git, the popular and widely used version control system for software development. It was created by Linus Torvalds in 2005 to manage the development of the Linux kernel, but it has since become a standard tool for managing source code across many different types of projects. Git allows developers to track changes made to their code over time, collaborate with other team members, and revert to earlier versions of their code if needed. It works by creating a repository that stores all versions of a project's files and tracks changes made to them over time. Some key features of Git include the following:

- **Distributed architecture**: Each developer has their own copy of the repository on their local machine, which they can work on independently before merging changes back into the main repository

- **Branching and merging**: Git allows developers to create branches or copies of the code base that can be worked on separately before being merged back into the main branch

- **Staging area**: Changes can be staged before being committed, allowing developers to review and selectively commit changes as needed

- **Collaboration tools**: Git provides tools for collaboration, such as pull requests and code reviews, that allow team members to review each other's work before merging it into the main branch

Migrating project code from an existing Git repo, either on-premises or SaaS, such as GitHub or Bitbucket, should be relatively trivial. Azure Repos supports the standard Git API and CLI tools, so whatever recipes or tools for source code repo migration will work for you. It is worth mentioning that there are some limitations, particularly in situations where you store very large files (50 GB) or binary files generally, even very small ones. Frequently committed files that contain many changes make the history difficult for Git to manage. However, it is very likely your teams have encountered issues already if this is the case, so it could be time to clean up your repos before migrating them.

Summary

Cloud migration is a process of transitioning applications, data, and other IT resources from on-premises to cloud environments. With cloud migration, organizations can leverage the scalability and cost savings that the cloud offers without having to invest in expensive hardware. The cloud also provides an opportunity for businesses to modernize their existing application infrastructure by using VMs and containers. By taking advantage of cloud services, organizations can unlock greater agility while reducing costs associated with traditional application deployment models.

In this chapter, we explored a number of common migration scenarios and tools that you can use to help assess your own situation. We discussed some cloud fundamentals on networking and compute resources and highlighted how these will differ from a more traditional IT environment. We also delved into application containerization, something that is likely not available in a traditional IT environment but may offer numerous advantages that could be worth the investment in training during the migration to the cloud. In the next chapter, we will introduce the concept of "cloud native" and how it can help you transform your organization to maximize cloud benefits and displace your legacy IT practices.

5
Becoming Cloud Native

Cloud native is a software development strategy that focuses on leveraging service-oriented technologies and modern software architectures to create and manage applications, with minimal operational overhead. This approach has become popular due to the obstacles associated with older monolithic app architectures, which are laborious to scale and maintain, often suffering from sluggish deployment and inflexibility in updates. Cloud native techniques provide enterprises with greater agility to respond quickly to market changes and increase innovation opportunities.

In this chapter, we're going to cover the following main topics:

- Cloud native development
- Architecture patterns
- Azure Platform as a Service (PaaS) enabling cloud native applications
- How cloud nativism supports organizational agility and innovation

Cloud native development

Listen up, folks – if you want to be in the top tier of tech, cloud native is where it's at. It's not up for debate. Whether you're a start-up launching a new service or a corporation trying to modernize an old one, cloud native is your only option. If you want to gain access to the latest and greatest tech and attract top-notch talent, you better believe that cloud native is the way to go. Trust us, you don't want to be left behind in the tech dust.

Okay, so you might ask, "*What do you mean by cloud native?*" That's a fair question, but the answer is (at least) twofold. Firstly, cloud native describes a software development approach that involves designing applications to take advantage of the flexibility, scalability, and resilience of the cloud infrastructure. This means you build your application for the cloud so that it is *native* to the cloud. Does that make sense? For example, a microservices architecture enables horizontal scaling (adding more servers), which works well on the cloud. The second part of the cloud native answer relates to which underlying technologies you will use to implement your architecture. This can be confusing for newbies and a point of contention between those with some experience. The **Cloud Native Computing Foundation**

(**CNCF** – `www.cncf.io`) promotes a vendor-neutral and largely open source technology stack for cloud native implementations. Conversely, many others (the authors included) view leveraging cloud **Platform as a Service (PaaS)** – that is, those services that remove the need to manage infrastructure (including the installation and configuration of software that runs on top of said infrastructure) – as the best fit for cloud native implementations, even with the potential cloud vendor lock-in.

Architecture

Cloud native architecture isn't anything new in most cases. It is just applying patterns and practices you already know and love, usually to extremes. High availability is a well-known concept with proven architectural patterns. For example, using multiple VMs behind a load balancer was the regular legacy pattern for high availability, but now that is done through multiple VMs, across availability sets, availability zones, and Azure regions.

High availability is usually a pattern that is applied if you need your service to stay up for an appropriate amount or percentage of time, depending on business needs. For completeness, let's define (quickly and loosely) some of the terms used in the previous paragraph:

- **High availability**: This means increasing the resiliency of hardware and/or software to single (or multiple) points of failure by taking different approaches and applying different patterns based on the business needs. Yes, it is true – not every service needs five 9s (i.e., 99.999%) of availability, usually because every single nine you add is prohibitively costly when compared to the previous nine. The usual high-level patterns are the duplication/multiplication of resources, asynchronous integration, the decoupling of services, and caching, but they can also be things such as deployment processes and regression testing that are usually (and erroneously) not considered high availability patterns – even though they are usually less costly than throwing infrastructure at a problem. If you have a poorly tested service, the usual high availability patterns may still help you but at the exponential cost of implementing some basic regression testing.

> **High availability but at what cost?**
>
> Without going into too much detail, just imagine the following scenario.
>
> You have a service that kills a VM every 30–60 minutes on average. Let's assume you didn't know that it does that because you haven't tested it properly, and you deploy that service to production.
>
> Without high availability, that service will start fine but will then fail. So, let's introduce some high availability patterns.
>
> Restart the machine when it fails. There is some downtime during the restart. OK, let's add a load balancer and deploy two VMs.
>
> Given the 30–60 minute window, they could still both fail at the same time. OK, let's deploy 10 VMs. You are now at 10 times your usual cost.

But let's not stop here. This service is needed globally, so you introduce it to four different customer regions across eight different Azure regions (four pairs of paired regions). We are now at 80 VMs globally. Not only do you pay unnecessarily for all these but you also pay the increased cost of operations for a team that manages this infrastructure, you get alarms for restarts and spin-ups of VMs, and you make noise for everyone in the system, who now need to cross-check every error with the restarts of the VMs to see whether that was the cause.

Let's also suppose that the service, when working correctly and not randomly dying every 30–60 minutes, can handle all the requests on just one VM. So, you deploy eight in total, one in each of the paired regions, and you pay less and don't annoy your developers, your operations team, and your customers (because, remember, customer requests are failing and have to be retried) – all in an effort to avoid properly testing your service.

At this point, you have to ask yourself, how crazy am I? The answer, of course, is very.

And worst of all, this isn't a theoretical issue. I've seen this issue in so many customers' environments due to their lax testing practices.

This example was just that – an example of high availability, and how there might be cheaper and easier alternatives to it. This was emphatically not an example of a cloud native pattern. We'll get to those quickly, I promise.

- **Availability sets**: These serve as protection against hardware failures by distributing the load across more than one identical VM. These are distributed across multiple fault domains (fault domains do not share power and networking).

- **Availability zones**: These are physically separated zones within an Azure region. Zones consist of multiple physical data centers, and they are separated and do not share power, cooling, and networking. This helps protect against data center-wide failures.

- **Azure regions**: These are regions around the globe that provide Azure services. They consist of multiple data centers.

- **Azure paired regions**: Azure regions that are paired are located in the same geography, and they are distant enough from each other such that a potential failure in one does not impact the other. In addition, Microsoft guarantees that software updates to Azure services will not happen to both paired regions at the same time, ensuring further resiliency. One consideration here is the latency, which is/may be higher, as these are geographically distributed regions, and the speed of light and other considerations apply.

Returning to the subject of cloud native, it is not just about taking your monolith (even if it performs well) and hosting it in the cloud (and maybe even wrapping it around or into a few cloud services). These services and applications (usually monoliths, but not necessarily always so) are built with all the constraints of on-premises architecture, on-premises hardware, and on-premises biases.

So, what are cloud native applications and/or services? Are they microservices? Are they Kubernetes-based? Are they highly available? Are they always better?

Whoa! Hold on there, partner. Let us try and answer these in reverse order.

Are they always better?

Cloud native is definitely a better approach to software development. Smaller, decoupled cloud native services are easier to maintain (although potentially, it is harder to operate a *"fleet"* of microservices) and can be used for rapid research and development. Cloud native allows the focus to be on business priorities, by leveraging cloud platform services that you would need to create, build, deploy, and maintain. It is a lot easier to deploy some code using Azure Functions than it is on a VM, with a CPU, RAM, OS, and software dependencies. So, when would this not be the best approach? Well, sometimes you might not need it. Perhaps you have a monolithic legacy software application that just about works, but is not worth the effort it would take to re-architect it. So, yes, cloud native is better, but you should still think whether you really need it, especially for existing services.

Are they highly available?

Focusing on cloud native allows you to realize the full benefits of the cloud, and one of those benefits is elasticity (the capability to scale); however, you don't always need high availability. For example, if you send a bill at the end of the month, you can get away with a service that isn't highly available. If any failures occur, you can always investigate, fix, and redeploy it again in a reasonable amount of time – and perhaps receiving the bill on the 1st or the 4th of next month means very little to folks. So, should you then invest in high availability for every service? No, not unless your landing zone has this built in – in which case you may avail it anyway, rather than make an exception for this one service, but only if it aligns with your business goals. Remember, cloud benefits include the following:

- Speed of deployment
- Automatic software updates to underlying services
- A shared security model where you focus on securing your apps (applications) and your infra (infrastructure), while the cloud provider does the same with underlying services
- Scalability/elasticity
- Essentially unlimited resources – compute, storage, networking, and so on
- Automated backup/restore

Elasticity is just one benefit that you might or might not need. While high availability isn't a requirement for your services, if you do need it, it is easier to achieve in the cloud, and it's even easier still in a cloud native-based approach to software design.

Are they Kubernetes (K8s)-based?

They can be, and a lot of people will tell you they should be, but they do not have to be. If you are still managing layers of abstraction over actual hardware, you are not cloud native (for example, managing the underlying infrastructure). If it requires a lot of effort and/or knowledge to maintain, it is not cloud native. If you use a K8s PaaS offering that completely mitigates you from knowing, understanding, caring, and/or managing the infrastructure, then yes, it is cloud native. However, in that case, why not go with an even better cloud native approach – serverless functions?

Are they microservices?

They certainly should be, yes. The only reservation here is the definition of microservices – when can a service be called a microservice? Some services will naturally be large (dare we say, *monolithic*), and that is okay if that is your approach. Microservices should fit the following definitions, and if you achieve those goals, who's to say you are doing it wrong:

- Decoupled
- Independently deployable
- Maintainable and testable (in both isolation and integrated with all other services)
- Observable
- Organized around a business function (or a domain)
- Owned by a small (two-pizza) team

To be agile, they should also be self-contained as a product and owned by the team that defines new improvements based on data, customer requirements, and future vision.

What are cloud native applications?

Finally, we get to answer this for you. Let's start by understanding what cloud native services are.

They are built from the ground up, optimized, and built to understand cloud concepts and services, such as cloud scale, cloud performance, cloud observability, cloud security, business agility, and, above all, the availing of managed services and continuous delivery, in order to deliver value for customers and the market.

Let's go into more detail and try and grasp some of these concepts.

The most important thing is the knowledge that the service you build will be cloud native. This might seem like a box-ticking decision, but it really isn't. You and the team need to understand what this means. Remember the patterns we discussed earlier? You need to understand that a cloud native service must be built in a different way than any pre-cloud (on-premise) service. Here are some examples:

- Developing and deploying a cloud native service is different from the usual on-premise applications. CI/CD is a must. Deploy to production as soon as possible. Why? Because the business will expect it. Gone are the days of long planning and market releases.

- For a service to be maintainable, it must be thoroughly tested because now there are millions, billions, and even trillions (yes, really) of requests coming at you every second. You cannot discover in production that your service doesn't work properly because the scale prohibits this. The best-case scenario would be that the service failed, but you won't get off so easily. If the service fails, it will have the following consequences:

  - Disappointed customers

  - Corrupted data

  - Disruption to your other services, with wrong data and excessive calls, or bringing them down outright

- Security for a public service is different than anything you've ever seen before. Security on-premises is (technically) doable by blocking everything from the outside, whereas security in the cloud is a separate topic on its own. Not only do you now have to cater to cloud native services but you also need to configure access security properly (such as not opening your storage account to the public), and you have to cater to a significant number of generated security events, data that is shared among customers, and the regional and local segregation of responsibilities.

- Previously, your customers may have been supported by you, and they would have had to wait for your team to wake up and fix an issue. Now, in the new cloud native world, you deal with a multitenant service that is globally available – your support now needs to be available 24/7, 365 days a year, and your observability must be at a whole different level than before. Previously, you could dig through logs and find information. You can still do that, but your **Service-Level Agreement** (**SLA**) is ticking down. Previously, you had GB of logs; now, you are lucky if you know what an exabyte is. Previously, customers could report issues; now if they do, your support team will be overwhelmed with tens of thousands of tickets – all for the same issue your service caused.

The good news is that cloud native design is based exactly around making that simpler for you. Previously, you needed to know how to deploy your web server, how to maintain it, and how to debug issues in it; now you need to know all about the cloud native services.

Services

So, which services are cloud native?

First, let's talk about **infrastructure as a service** (**IaaS**) and PaaS.

IaaS provides underlying services such as the following, which you would expect anywhere:

- Compute – for example, VMs

- Networking

- Storage

PaaS provides services such as the following:

- **Azure Functions**: A service to host (snippets of) your code without worrying about or managing the underlying hardware, networking, or storage (you still might need to configure some of these with a light touch, but it is a *configure-and-forget* type of approach).

- **Service Bus**: A service to send messages between your services (thus, essentially decoupling them – if you take care of versioning, etc.) in a resilient way (with messages backed up, distributed across regions, etc.). Keep in mind that you know nothing about the hardware you run this on.

- **Anomaly Detector**: A service that allows you to easily start detecting anomalies in your services by just ingesting time series information.

- **Azure SQL**: A service to host your SQL data on without installing SQL Server and managing it.

- **Azure Active Directory**: A service to manage single sign-on (among other things) for employees or customers, or to integrate single sign-on with other apps and companies.

- **Media Encoding**: A service to (you guessed it) encode media. This is available at scale and in a few minutes, with no need to know how exactly it works and what hardware it uses.

What is my favorite Azure service?

The Azure SignalR service. This is a service to add real-time functionality to your application, and one service that is very underutilized as a cloud native service. Everyone uses it for chat (especially in demos), although it is surprisingly useful for any real-time communication, such as between frontend and backend services, as well.

SignalR can significantly improve the cost and latency of multiple requests for the same information in, for example, an e-commerce use case. Another example would be when you distribute information that keeps changing often to a fleet of services; clients can use SignalR to get real-time updates, rather than individually requesting this information. Another example is if you call many different services in an attempt to find the best one, where you can have them all join SignalR and let them compete, rather than calling each of them, paying for it, and then deciding which one to use.

These are simplified use cases, but I am trying to get you to think about what is possible and how you may mistakenly think you are working at your most efficient when, in fact, cloud native has an ace up its sleeve – SignalR!

If you want to get the most from the cloud, you must choose the right tools for the job in order to be productive throughout your service life cycle. Business, management, developers, operations, and security – all are impacted, and in order to get the best from the cloud, you must take them on this journey with you.

When building and deploying microservices in the cloud, you can choose between serverless options such as Azure Functions or container options such as Docker. However, only the optimal combination of services will yield the desired results. Your primary focus should be on serverless solutions, which have the added benefit of reducing infrastructure costs. By doing so, you can concentrate on your company's core business rather than cloud management. We recommend utilizing fully managed serverless solutions or exploring this option before considering other routes.

Another short story

Recently, I architected a solution for (a lot of) third-party integrations. The approach suggested was taking a part of a monolith service and containerizing it. However, that brought costs, scale, and rework considerations that needed to be considered.

What the team eventually built was a completely serverless solution with less than 10 functions altogether, and with an easy path to integrate further third-party services. The best thing was the low cost! Landing zone services (load balancing, networking, storage, etc.), Azure API Management, Azure Functions, and Azure Service bus combined amounted to just a few dozen euros, rather than thousands. Scalability? Built in. Observability? Built in. High availability? Built in. Maintenance? It was so much easier, focusing on one tiny function at a time.

To show you how easy it is to build a service entirely based around Azure PaaS offerings (meaning you don't need to manage infrastructure, even though sometimes you need to make a few choices around the size and scale of underlying compute, memory, and networking), here are (some of) your choices on Azure.

Compute

- **Azure Kubernetes Service**: This is for all your wonderful microservices packaged as containers. This service supports blue/green deployments, container versioning, and scale elasticity. As always, security is a shared responsibility, but Kubernetes comes with a steep learning curve. Note that this can be made serverless by using virtual nodes under the hood.

- **Azure Functions**: This is for all the mini-code snippets that will run over and over and over again, with endless elasticity at your disposal. This is *the* service if you and your organization are smart enough to focus on your business functionality.

Tip

I start any architecture (and so should you) by thinking about what services can be used to deliver the organization's business case to fruition. Then, I think about the Azure SignalR service and how I could leverage it in every solution.

I once discussed with a customer a migration of three data centers with 50,000 VMs in each, and my first thought was, how can we do this in a serverless way? And you know what? A lot of VMs were migrated to cloud VMs, but an awful lot were modernized and deployed using Azure App Service or refactored as Azure Functions, and – yes – Azure SignalR was in the mix as well.

- **Azure App Service/Azure Container Apps**: Want to host web services? Feel that Kubernetes is overkill? Azure Functions is messy and sprawling all over the place? I might disagree with one of the last two questions. However, you may prefer one of these options. Azure App Service supports many languages natively, and also containers – so take your pick. Websites are perfect for this service. If your marketing department needs a place for them to manage a million landing pages or different microsites, you could do worse than this service.

Data storage

- **Azure Blob Storage**: This is storage. It's infinite (for all practical purposes). Usually, it's not thought of as serverless, but really, there are no servers for you to manage.

- **Azure SQL**: This is for all your relational needs. You need to consider whether what you are doing is the best way to scale and be truly relational, or whether you are just forcing everything into the most expensive storage available. Also, we are now in the world of the cloud and cloud native. If you licensed a database on-premise, it (kind of) makes sense to stick with it throughout, but now you are in the cloud. Some data fits well in a relational database, whereas some does not. So, why don't you have both?

- **Azure CosmosDB**: Speaking of another amazing option, welcome to CosmosDB! This is your default database. CosmosDB should be your first choice, and only if it really isn't suitable should you consider others. The simplicity of a serverless, hyper-scalable database with the four SLAs (*availability*, *throughput*, *consistency*, and *latency*) is unmatched. It truly is a galactic-scale database. I fully expect it to be replicable to other planets and spaceships in due course – as it is today across the globe – with literally two clicks.

Workflows and integration

- **Azure API Management**: Do you need to serve APIs? Of course you do! What kind of cloud native architecture has no need for APIs? I know you will want to tell me; I am looking forward to your feedback.

  Internal and external APIs should be the same, and the Azure API Management service allows you to not only host them and package them as you want but also to provide a self-service portal, for registration and testing against your business APIs. It also scales and allows throttling (which you should do on all published APIs and endpoints) and monetization should you desire it. Remember that consumption-based services are to cloud native what food is to humans – essential.

- **Azure Logic Apps**: Cloud native isn't easy to pull off. A huge disadvantage (that, nevertheless, does not negate the advantages – not by a long shot) is the orchestration. With cloud native, you will have quite a few services all over the place, and if your organization doesn't take the digital transformation mandate and the Cloud Adoption Framework seriously and (re)organize itself to be able to take advantage of this, there will only be pain. Beware – your cloud native architecture means nothing if your organization isn't able to deliver on it. A service that can help connect a lot of these services is the Azure Logic Apps service. Whether done by developers or citizen developers, it allows relatively easy integration of your services among themselves, with external services added to the mix as well. Sometimes, you will need to fall back to Azure Functions, but always try and consider whether you can do your integration in Logic Apps. If you haven't already, it is very easy to try out. Its drag and drop functionality and a marketplace of already existing components will make you a pro within days. And think of your customers as well – when they see you for the first time, integrating their services with yours within minutes (days and weeks should then be spent on testing and ensuring fallback, degradation, and edge cases are covered), you will sell more software than ever before.

Want to WOW them? Here's how...

Another amazing way to sell your organization's services is to have a *live* architectural diagram as an app and allow customers to pick and choose, letting them kill live any service they want – *live* in production. They are always blown away by that. Beware, again – this, you can only do if you have your cloud house in order , you understand the Cloud Adoption Framework (which you have prepared with elasticity, high availability, and graceful degradation patterns), and your organization has all the observability it needs. A live architectural will look cool – on stage or in a board room, as nothing screams *we know what we are doing* louder than allowing a customer to destroy your live environment without any consequences. However, be cautious – test, test, test, and test again before you let the chaos monkeys (which is a fun (above all else) pattern; look it up) loose.

- **Azure Service Bus**: Decoupling is a word many take lightly. The first service AWS launched was S3 (storage). The second one was Amazon Simple Queue Service. That is how important decoupling is, and for cloud native, it is essential. Azure Service Bus is an equivalent service from Microsoft (there are others with this functionality, but this one is essential). Decoupling allows you to infinitely scale your development efforts, and queues and topics are the tools to achieve that. You want to ensure that all the services your organization makes are independently deployable. A great way to achieve this is by decoupling your services through asynchronous messaging rather than synchronous API calls. Why is that so useful? As long as the message format is well defined and supports versioning, your services couldn't care less what other services are doing, or how they are evolving.This just means either that messages are static and services always send them the same ones, even when the service business logic changes, or that services are aware that messages are versioned and are agreed across the board to support all versions. During the lifetime of your overall architecture, both of these statements will be correct most of the time, as (and this is a general principle that isn't necessarily always correct) messages shouldn't change often, but when you need to make changes, it all still works. Relating to just that last sentence alone, there is a body of knowledge out there you can consult if you want to know more – and you really should want to know more. Remember that queues and topics are essential to cloud native.

There is (at least) another popular way to achieve decoupling, but if you start this way, you will never want to do anything else – unless there is an overwhelmingly compelling reason.

Observability

- **Azure Monitor with App Insights**: How can you know whether your services perform well or at all if you don't have observability sorted? If you have a few services or a monolith, you can log and grep (a command or a utility to search through logs (and other files)). If you have 10, 100, or 1,000 services storing logs all over the place, searching through them is next to impossible. Luckily, there is the Azure Monitor service and the wonderful **Kusto Query Language (KQL)**, which make this a piece of cake – *as long as you send all telemetry, always and consistently!*

Analytics and AI

- **Azure Stream Analytics**: This is a service that enables you to ingest and act on a truly humongous set of data. Whatever your organization does, it must gather data. It should responsibly take care of it, but data is oil, data is gold, and that very specific data is the one thing your company has that no other company has. Can you really afford to throw out such an advantage? No, of course not. Azure Stream Analytics can come in really handy here.

> ### Mining the cloud
>
> Two of my customers were companies that ingested telemetry from dozens of car manufacturers – from each and every car sold – from the moment they rolled off the assembly line until the moment they were crushed or disassembled. We are talking GBs a second.
>
> Now, every sensible person I ever spoke with would say, *"Don't ingest all that data; just sample it."* Right? WRONG!
>
> Data is the new black. Data is the new oil. Data is what will remain once money, oil, and whatever replaces them (if we manage not to destroy ourselves in the meantime) have long gone. Data is essential when making decisions, and I expect you to gather data and check it twice before you apply any of the advice in this book in a cloud adoption, or before you commit to an architecture solution. So, you want to ingest all the data, and you (with your entire organization) can then, and only then, decide how long to keep each data point.

- **Azure Cognitive Services**: Sometimes, you need to create your own models, manage them in production, check and manage drift, and deploy others. And sometimes, you can use an existing model that is managed for you through a managed service with a few APIs you can call. If so, check out Azure Cognitive Services first if you have a need for speech, language, vision, decision, and should be other things that you will surely be dealing with.

CI/CD

- **Azure DevOps**: If you can't deliver value to customers every day, you might as well pack it in. No joke. Stop what you are doing, as it isn't good enough. OK, it was a joke. But think about it – really think about it. Alternatively, get better and start narrowing the gap between developing a service and placing it in the hands of your customers (remember, they might be internal as well). We'll get to agility quite soon in this chapter, but for now, remember that Azure DevOps is a service that you must master and use fully to deliver value to your customers on a daily basis. Building and deploying pipelines (as code) is crucial, and all the other services in Azure DevOps are there to support these two things. You need to build your code and deploy your code – often, automatically, at speed, and with no manual gates. Nothing else will do in today's environment.

You can use all the preceding services without ever logging into a server. To conclude the cloud native introduction in this chapter, you must do the following:

- Build cloud native services, as nothing else will suffice (you have invested in the cloud, and to get the most value out of your investment, cloud native is the only way to go).

- Know how to choose which service to use and when and why. (The good news is that picking the wrong service isn't terrible, as you can spend some time before you realize you made the wrong choice and try to correct it. Of course, it is still better to be correct most of the time, and if you go with the services we just discussed, you can hardly go wrong).

- Prepare to manage the chaos that is 10, 100, 1,000, 10,000, or more services. Since this is your future, prepare for it in the present. Once a business sees how agile you are, they will want – nay, demand – more and more services.

- Integrate these services and teams into an architecture and organization that knows how to execute quickly and efficiently (your competitors are doing it as well!).

- Do CI/CD well and automate everything. Without automation, there are no SLAs, no daily deployments, no high availability, and no research and development; there is only pain for your development, operations, and security efforts.

- Ensure compliance, security, reliability, cost optimization, operational excellence, and performance efficiency (if one is not in the mix, you will have terrible days ahead of you – anything from ransomware and data breaches to lateness to market and developer and customer churn).

- Organize your cross-functional teams to distribute accountability (an oxymoron if ever there was one) and actually deliver on your business and architectural objectives.

- Control (or attempt to as best as you can) the creative chaos of business demands and the efficiency of delivering services to meet elastic demand. The best pattern of control is actually to relinquish control – everyone in an organization must be at their very best and ready to step in and deliver on what is asked from them by the market, customers, and the business objectives themselves – the honest truth is there can be no control from a single tower, but if each person can erect their own tower when needed to accomplish a goal and everyone else is happy to climb up and help you, then you are well on your way to being the best you can be.

Your job – yes, your job – is to enable an organization to understand what needs to happen and what needs to change, and this is a never-ending job. Good luck. The rest of us can attempt to help you through books such as this one and the patterns and practices we promote.

Agility

Agility is simple. Forget about consultants and frameworks because they are mostly useless. How can things such as a consultant or a framework help you be agile – is everyone the same, is your organization the same as the next one, and does a consultant or a framework know your organization and have the ability to help it? Accept only consultants that are there to point out issues you have and then let you pick the solution to fix the problem, as only you know the organization's culture, people, and processes.

If there was a framework to build the best football team, there would be one. But there isn't and there will never be one. Is 4-4-2 the best formation? Why not? The answer is because of the different cultures, people, and processes at each football club.

Agility is not SCRUM, SAFe, or another framework.

All you must do is ask yourself a few questions to see whether a framework helps you:

- Are you delivering value at least daily?

- Are you listening to your customers (internal and external) and delivering value to them as fast as possible, at least slightly faster than they require?

- Are your teams happy and working well, as well as sharing your customers' values?

If the answer to the preceding questions is yes, then the process you have is great. Double down. If that process is SCRUM or SAFe, excellent. Well, not excellent really, but you get the point hopefully. ☺

If, however, you cannot answer all three questions positively, then no framework will help you, and you need to focus on changing your organization to be able to answer yes to these three questions. Yes, you can bring a consultant in if you lack experience or if you want an external perspective. Remember that they are there to point out your issues, but *you* are the only one to solve them. From the time your organization innovates and thinks of a big idea to the time it delivers – is it days, weeks, months, or quarters, or worse (or a lot worse)? Can you deliver rapid innovation, or can't you? There are always why nots – why things can't be improved. Your job (and everyone in the organization) is to make it happen and enable everyone else to make it happen by whatever it takes to do so; otherwise, you are not doing right by your organization.

Let's talk about the *why* for a few paragraphs, such as why agile organizations in the cloud can deliver rapid innovation. It's because only with the combination of the patterns and practices described previously can you deliver rapid innovation – innovation every day, small, large, and huge, across the entirety of your business, across functions. This includes innovation with customers and partners, and innovation that delivers value, then more value, and then even more. This can only happen if services are aligned, if an organization is aligned to innovate, and if achieving agility itself is delivered in an agile way.

Who doesn't love recursion? Not only must you deliver rapid innovation in an agile way, but you must also keep in mind that things that worked last time (a week ago or a month ago) may no longer work, so you must innovate on your innovation process, and how you enable agility must be refined and innovated too. What worked yesterday may not work today!

Think of your tools and processes. Check Azure services and see whether they can help you. Check the industry practices (*not* the frameworks) and evaluate what will and won't work for you.

Hire the best people – even diverse, unorthodox, and downright megalomaniacal people. Then, allow them to be their authentic best selves. Did you hire the best? Then let them do their best! Are your processes in the way? Innovate. Quickly! Only then will you achieve true agility and true rapid innovation.

Innovation

Leaking in from the previous few paragraphs on agility, there is innovation. You know now why it is important and what you must do to deliver it. But how can Azure assist?

As mentioned, this process of yours regarding rapid innovation must rapidly innovate itself, but here are some examples from Azure on how to get closer to achieving it:

- Remember decoupling via Azure Service Bus? Well, by using topics (a type of message queue that delivers a copy of a message to each consumer/subscriber), you already enable future innovation you don't even know about. How? Well, today you send a message to the topic, and it is processed by the consumers you know about, but tomorrow, there will be other services interested in the same message, and all they need to do is subscribe and consume the message without any other change to your architecture or services. So, if someone has an innovation workshop and comes up with a new service, they can, within hours or days, test it with the copies of messages already flowing about. This is a powerful enabler of innovation.

- API Management is another way for an organization to own the developer experience. Cataloging and publishing your APIs is an easy way to open up rapid innovation. The more services you have, the more valuable APIs you have – and the more innovation you can foster. New services can start to use existing APIs to do more thorough integration with existing services. Just having visibility across all the APIs of an organization fosters innovative thinking if you've enabled your people to stop, learn, think, and innovate.

- Logic Apps is a great tool to prototype and test new integrations, to see how new services work together, and to smoke-test for the viability of investing more in an idea. Anyone can drag and drop, and with a little support, everyone in your organization can be a citizen developer and try new things, especially if your organization has an API catalog available to all.

- Allow Azure Monitor access to an entire organization, together with all the dashboards, metrics, logs, and alerts. Let everyone in the organization have access to these, and let them create their own queries, searches, filters, dashboards, and alerts.

> **Use data – don't assume…**
>
> I always expect (demand?) that an entire organization is on the same page. All of us should know dashboards by heart. This is especially true for start-ups! If you don't know how your services and, consequently, your customers are doing and how they use your systems, how can you make day-to-day business decisions? How can the marketing department do its job, how can architecture do theirs, how can compliance do theirs, as well as security, engineering, customer success, and finance? How can anyone in the organization do their job if they are unaware of what customers are doing, where your solution is failing to meet their needs, and where it is succeeding? For an enterprise company, this is a background task to have in mind, to regularly check and double-check assumptions. For a start-up, this should be a day-to-day activity!

- Azure Cognitive Services is improving, and new services and capabilities are being added. This is also true for the Azure catalog of all services. So, in order to innovate, sometimes you just need a service to become available, and then something clicks in your mind on how this is a great service to include with your own services, adding even more value to them. Keep this in mind. No architecture is complete or so perfect that you shouldn't revisit it occasionally, and one trigger of this can be a new service announced by Azure.

Another important guideline is to listen to and address feedback from your teams, and keep improving your governance and processes to allow for more agility and innovation. Everyone should innovate, and everyone should propose and try things. With cloud native, agility, and innovation, there will be plenty of innovations – some may stick and become new services or new processes, some may improve slightly on the existing ones, and some you can discard after giving them due consideration. You must be ready to fail, fail fast, and fail often to arrive at those innovations worth pursuing. Embrace failure, and you will learn a lot. Sometimes, you will learn what to do, and sometimes what not to do.

Explore opportunities to drive innovation within an organization through greater flexibility, openness, and agility. In doing so, you will have happy people, happy customers, and a happy business.

Summary

With cloud native technologies and practices, organizations can unlock new levels of flexibility, openness, and agility, driving innovation throughout their operations. Whether through the use of tools such as Logic Apps and Azure Monitor, or by taking advantage of new services and capabilities as they become available, cloud native technologies provide organizations with the tools they need to constantly explore new ideas and improve their systems. Thus, by embracing cloud native practices and technologies, organizations can stay at the cutting edge of innovation within their industry.

In the next chapter, we will explore how you can prepare (or likely transform) your organization to be a digital cloud native.

6

Transforming Your Organization

Don't underestimate the number of organizational changes required to be able to deliver on your goals of adopting the cloud. There is a huge difference between using the cloud as you would use any other technology or set of technologies and adopting the cloud and becoming the kind of organization that can truly deliver on its objectives by becoming a cloud-native organization.

In this chapter, we will cover the following main topics:

- Organizational transformation
- Site reliability engineering
- Cloud governance
- Communication planning

Organizational transformation

A key organizational differentiator today is agility, or more specifically, an organization's ability to react and adapt to change – change in the market, technology, and people. Agility enables an organization to adopt the cloud, utilize it effectively, and gain the most benefits from it.

Anyone can deploy workloads into the cloud. But you need organizational transformation to be able to deploy workloads into it with the benefits of speed, efficiency, cost optimization, security, and so on.

What's the difference between those that truly get the cloud and those that don't?

Imagine for a moment two different organizations. Organization A has one product, deployed on premises to hundreds of customers. Organization B has tens of thousands of applications (internal and external) deployed in three of its own data centers (across 80,000 VMs). The teams in both organizations are equally skilled and equally capable of working with cloud technologies.

Both organizations claim that from January 1, 2023, they will be cloud-first and cloud-only – so they both start planning for the move. Six months later, organization B has moved 80% of all their applications, closed two data centers and by the end of the year is on track to get the final 20% of applications across. Two years later, organization A has not moved a single customer to the cloud.

Both leadership teams still claim they are a cloud-first organization – to my face. What is the difference? One organization has ignored the advice about the need for organizational change and has remained a bureaucratic, top down, hierarchical structure with claims of scaled agile practices and agile coaches (while there are very good agile coaches out there, I would trust my teams to organize themselves rather than enforce top-down "coaching" from the sidelines – so be careful selecting the coaches) – and the other organization has transformed into a modern agile powerhouse.

The only reason for such huge differences between these organizations is that the leadership team in one has truly understood and adopted the cloud, while the leadership team in the other has not understood and therefore not adopted the cloud, but they are happy to pretend they have and when doing an analysis of why they haven't moved, they found causes in market positioning, commercial proposition, team skills gaps, partner readiness, and team maturity. However, the simple truth is that without an organization transformation that enables agility, parallelism, and decision making based on data and customer success, they were doomed from the start.

I've had both organizations as customers. I spent 99% of my time with organization A being ignored, waiting for action, and in meetings. I spent 99% of my time with organization B advising, engaging, and unblocking actions across teams, with individuals and even with their customers.

Teams in organization A had retrospectives after every 2-week sprint and delivered almost nothing of value each time, and every decision was made by a committee of people from all levels of the organization. Teams in organization B, on the other hand, didn't even have defined sprint lengths, there were no retrospectives – but each team dealt with any issue swiftly and efficiently across all levels of the organization. One of these organizations transformed, the other pretended to. Only one of them is now running a true SaaS platform, to the delight of their customers. The customers of the other organization are begging them to speed up their move to the cloud and are jumping ship to competitors' offerings or even their own. And you know you've made a mess when your customers would rather build their own thing from scratch than work with you or use your services. Very rarely is this a good thing for you.

So, what does it take to transform an organization and how do we achieve it? Let's find out:

- The first piece of bad news is that without C-level management understanding why the organization needs this change – without their 100% unequivocal support and desire to drive this change, there is no chance for you. The primary reason behind this is that without the budget, the drive, and the trust from C-level, this will be a tire-kicking, plate-spinning exercise, which will lead you nowhere. This is either an all-in push where, yes, there will be issues, failures, misunderstandings, and missteps, but there is no going back – we will adapt and go forward and eventually we will make it – or, this is an attempt that will hit the first few roadblocks and collapse.

- The second piece of bad news is that, despite most advice you find online (blogs/YouTube), there is no pattern you can wholesale adopt and be sure of its success. There are a lot of ideas and a lot of suggestions that have worked for others, but you must find your own way. Your organization has its own culture, people, goals, diversity, markets, commercials, and so on, so you cannot just do the same thing as everyone else and succeed outright.

- The third bit of bad news is that this is all on you. There must be someone to persistently lead the charge, point out when things stall, and encourage and educate everyone else. If you can find more folks like yourself, great. If not, then your goal is to make them. Because if there is no driving force behind all this effort, transformation will get bogged down in indecision.

There is good news as well. As this is a process, mistakes can and will happen, but learning from the experience is valuable and if you and the organization fail-fast and recognize this process for what it is, you can always course correct.

The longer the process goes on for, the more and more educated people are on it and as the culture slowly shifts, there will be more and more allies you will pick up along the way.

Honesty is your number one policy. Everyone in the organization should go into this process knowing two things:

1. If the organization is to do this, everyone must understand that this transformation is the goal and that you will not stop until you reach the peaks of agility that will enable you to execute your business vision across the entire organization.

2. To achieve number 1, every single role in the organization will have to change in order to adopt the mindset of continuous improvement and focus on the delivery of the business goals – not in isolation but together with everyone else.

Feedback and feedforward are cultural values that will need to be adopted. Similar to blameless post-mortems (we will go into detail about these later in the chapter) of incidents in your cloud environments, a cultural shift is needed around giving and accepting honest feedback on what works and what does not work in the organization and fixing those issues as quickly as possible.

This must work across the organization, across levels, skipping levels – up, down, side to side, and including contractors, partners, and customers. Of course, how you deliver feedback is as important as what the feedback is, and there are management principles you can apply, but generally what works be clear and concise, try not to criticize anyone. Requesting feedback is easy. Acting on it is sometimes difficult. So, always have a checkpoint to see if it has been acted upon and, if not, then why not?

People will be happy to give and accept feedback when they feel it is well intentioned, genuinely welcomed, and is then acted upon. Nothing will discourage people more from giving feedback than not acting on it.

Generally, if a piece of feedback has been delivered to you three times from a single person (usually with different wording), you will stop receiving it. Once you realize the feedback is not forthcoming, you have failed. There is no chance to improve without honest feedback across an organization.

If the culture is such that it discourages feedback, then everything will go on, to use the dreaded phrase, *as it has always been done.*

Forget technology, management practices, and business goals. If feedback is discouraged or blatantly rejected , you won't improve anything until you change the culture. The importance of building a healthy level of psychological safety as a pre-condition to offer and receive feedback cannot be overstated! It just means if you have a great workplace – a trusting and comfortable workplace – you will get feedback. If you do not, you will not. And if you do not get it, how will you ever improve?

Before we go into how to adapt and adopt new processes, think about why culture is important. Firstly, your organization is different to any others, so you cannot just wholesale apply the practices of another organization. Your organization is special, it has survived in the market, and there is huge value in it. But now you need to transition to the next level, and you can only do so with everyone pulling in the same direction.

How to adapt the Cloud Adoption Framework for your organization and vice versa

Before you can adapt a process, you must know the process – and make no mistake, you will have to adapt it.

The meaning of adopt is to *take something as your own*. This is what the Cloud Adoption Framework is – a framework for you to take and use. Not wholesale, and not indiscriminately, but with vigor and with a plan.

Adapt means to *change for a new situation*. Think of this as the workloads, the people, and the processes you can adapt in your organization.

You shouldn't and you mustn't try and change the Cloud Adoption Framework to conform to your organization so that the organization can continue to do as it has always done, but now with the added claim of *"we are cloud-first, cloud-native, and cloud-ready!"*

Of course, there are some parts of the Cloud Adoption Framework you won't need, won't want, or will dislike, or that (you think) have been proven incorrect. Granted, yes, there are always exceptions. But an exception is defined as *a special case*, which is either special or unique.

So, if you or anyone in your organization finds this special case, fair enough. But, firstly, it needs to be double and triple-checked regarding how truthful it is, and secondly, it should still be an exception, and everyone should be aware that it is an exception.

For example, your organization may decide that Red/Blue or Purple teams are not necessary for your SaaS platform as you have a partner that does these sorts of activities for you. So that is one adaptation you can make! Right? No. Or, more accurately, it depends.

The intention behind creating Red/Blue or Purple teams is that there are dedicated resources working around the clock, 24/7 on security testing. So, if that is the agreement you have with the partner and your other teams will stay in touch with them throughout their workday, then absolutely, this arrangement can work. So long as you drop in the partner teams as a like-for-like substitution for your staff, meaning for all intents and purposes, the partner security team is treated like an internal team.

If you, however, pay for quarterly penetration testing of your entire SaaS platform, that is not an acceptable substitution as it replaces 24/7 active engagement with passive once-a-quarter engagement.

The same goes for deviating from the well-established architectural patterns. As a very general example, *decoupling* is a very useful pattern in SaaS architecture. Deviating from this pattern is perfectly fine as long as the reasons are well understood, documented, and agreed upon. However, deviations (or design exceptions) have drawbacks:

- Exceptions can annoy you (and the people who need to implement them), in the back of your mind. That is how you know it is an exception. As it is a necessary exception, ask yourself if you can be annoyed and still live with it for now.

- If it is annoying and you can't live with it, it means it is the wrong solution, so you should try to fix it.

- If it is not annoying, then it is not an exception, which means you should try to fix it faster.

So, how do you rethink old ways of working? The cloud is the new normal, but the cloud isn't fully understood by all yet. Well, it is all about education. But to successfully educate, you must have willingness from the participants. Again, before doing any business or technical work, culture must be conducive to education, to change and to feed back. Is the culture in your organization like that?

As an architect, your job is to look at the high-level business requirements and the products, services, and technologies you are availing of today. Then, and only then, should you check out the Azure Cloud Adoption Framework site to get a high-level overview of what it entails to adopt the cloud. Just remember, Microsoft primarily deals with enterprise companies, so the solutions may not apply outright.

For example, enterprise connectivity between on-premises, your organization's locations, data centers, and the cloud(s) applies only if you have such requirements.

For a company born in the cloud (like most tech start-ups in the last decade), this may be a moot point.

For example, migration plans may or may not apply to your organization, depending on whether the organization has something worth migrating.

Do also think about who is going to be accountable for each workload. Is it you? If not (and it shouldn't be – unless you have only one workload to worry about), then who is? This again brings up the adaptation of the organizational structure to handle this.

Is Product Management a silo, Software Engineering a silo, Architecture a silo, the UX team a silo, Operations a silo, Security a silo, and is the Platform team a silo? Yeah, that's not going to work.

And I know that some of you will still want to try. Just place a bookmark here and come back after you've tried.

The only thing that consistently works is multiple cross-functional teams that fully own their product or service(s). And I know in the world of scrum, agile, scaled agile, and the hierarchical organization built up over years and years in your organization, you've done things differently. Number one, you shouldn't have and number two, you won't make the same mistakes from now on!!

I also know that some of you will still want to try. Just place a bookmark here and come back after you've tried. And yes, this is the second time now. Hopefully, there won't be a third.

There can always be exceptions; for example, if you look at the Azure portal, you will see (with minor exceptions) consistency across thousands of screens (portal blades) and services. This might be created by a single team due to the UI/UX consistency required. But it is, altogether, an exception.

So, who will be accountable for each workload? This is equally important to answer for workloads such as the migration of three data centers into the cloud as it is for a business-critical service such as health data record access and storage.

Now, you must ensure all teams are aligned and that they understand each of the following:

- **Business goals**: They are important because if you want to build a service, you need to know what the business is trying to achieve. Sometimes the answer is just something such as improved performance, but usually there is a further business reason behind the service request. Know it, understand it, grok it, internalize it, and think of the customer. With the customer in mind, you can start designing the high-level service.

- **Landing zone**: This is something you will need to build. Your services are not in a vacuum, nor will you be reinventing the wheel with each and every one of them. For example, scale can and will be dealt with all services in mind – security as well. The reason is simple really. All services should scale based on the load. All services should be equally well protected. So don't leave it to different implementations done by each team.

- **Architectural strategy**: Here is where the overall strategy comes into play. Knowledge that exists within the teams and the education needed to get them to the target architecture will dictate the actual target architecture. There is no use selecting, for example, Kubernetes as your orchestrator/platform if no one in your organization can secure and manage the thing. Either the knowledge must be there or extensive education must be provided, which will dictate the architecture. Maybe a better target architecture approach is a variant of serverless Azure Functions, API Management, and decoupled messaging through Service Bus.

Who will lead the landing zone creation? Who will create and inform people of the business goals? These are important questions. As you know, assuming breach and zero trust are some security aspects/patterns recommended when architecting services.

Who will ensure security, test, check, enforce and report on security incidents? If (when) you are breached, who will be responsible for executing the response? What is the response?

You, as an individual contributor, do not scale. Others need to be orchestrating as well. You all need to plan how you and the organization will orchestrate the work efficiently.

Don't be a know-it-all; be a learn-it-all. Delivery and operations must flow like spice (a Dune reference; we are huge fans of the books). Cloud services are numerous, as are cloud concepts. No one knows everything (except the authors of this book, of course!). So, you need to plan to map teams to workloads, team members to teams, and skills and learning paths and certifications to team members. You also need to plan for rotation (for example, in Blue/Red/Purple teams). You need to plan for escalations, communication responsibilities, and so on. There is a lot to unpack here. But the good news is there are existing frameworks and organizational patterns to help make sense of it all.

Site reliability engineering

In every organization, there comes a time to create new roles and/or realign the new roles and the old roles and teams and what their tasks are going forward.

Running services at scale is hard and hence a new way of doing things was needed. Meet **Site Reliability Engineering** (**SRE**).

Google has kindly provided a lot of free information on this in terms of (at the time of writing) three books on SRE. Find them at `sre.google/books`.

In this section, we will focus on the transition into the SRE way of thinking and why it is important to consider it.

The reason why this is covered in this book is to reinforce the necessity to pause and think about your cloud adoption. The Cloud Adoption Framework gives you the insight and the tools to get it done, but only if you have stopped and thought through the consequences of being a cloud-first organization.

The cloud really is different

I cannot stress this enough: adopting the cloud is absolutely not the same as doing the same things as before under a different name. It is a completely new way of doing things and with a completely new set of issues and possibilities.

As one example, in traditional on-premises IT infrastructure (or a private data center), you almost never have to think about noisy neighbors – people and workloads using the same machines or machines near each other. In the public cloud, this is an issue that you need to understand and plan your deployments around.

As another example in the cloud, you also must think a lot harder about which sizes of services you provision and why. It is usually a lot better to have thousands of smaller instances than a single large one. In traditional IT, it is a lot easier to manage one large database than it is to manage a sharded database across hundreds of geographically distributed machines. In the cloud, the opposite is true (to an extent), but also necessary. At cloud scales, your service may become popular quickly, by design or accidentally, so you must be ready.

SRE! Do you need it? Must you? Should you? Can you?

SRE culture itself is a much broader topic, but here we cover the absolute musts you can take from SRE culture and give you a way to decide when you really must invest more in this topic.

Story time!

Traditionally, there were developers and there was IT (operations).

Developers wrote code for some features and the wonderful products eventually made their way to customers and then, sometimes, there were updates. So, every year or so, the developer packaged their code and threw it over the wall.

On the other side of the wall was IT, who now needed to figure out how to deploy this thing, where to deploy this thing, and how.

If the organization was great, there would be a wonderful thing called documentation. And if the organization was exceptionally great, the documentation would be accurate and helpful. IT would than *make it work* and the code was in the hands of the customers after extensive testing. Then, a number of customers, ranging from 2 to 2,000, would access the product and try it.

Twice a year, there would be an update – either a missing feature would be added or an existing one would be fixed and it'd be thrown over the wall again.

On the other side of the wall was IT, who now needed to figure out how to deploy this thing, where to deploy this thing, and of course how while not interfering with the customers. So, downtime was scheduled and communicated. But then something strange happened. Bad people, or customers as they are traditionally called, started demanding more updates, more features, and more reliability.

A yearly release cadence was not acceptable to customers. A fix in the next quarter was not acceptable. Downtime for two days was also not acceptable. At the same time, the other lot of bad people, or sales as we call them, started selling more and more services to more and more customers. So, it wasn't 2 customers anymore, or even 20, 200, or 2,000; there were now 2,000,000,000 customers. They all had data, they all had tasks, and they all needed data safe and tasks completed. Think, for example, of services such as an email service or a web search service. The sales and the business teams complained to the engineering team that if they could think of and sell these services quickly, why couldn't they have the service faster and more reliably than before?

Among the first to realize this were the people of Google, who just so happened to be running web search and email services for billions of people. And that's how SRE was born.

SRE is what you get when you ask a software developer to do operations. Here is what changed:

- Gone are manual deployments

- Gone are walls

- Gone are unnecessary customizations and distractions

- Gone are one-year cycles

- CI/CD was introduced

- The SRE team does both development and operations (40:60 or 50:50 being the best mix)

- Standardization and governance were introduced

- Cycles are now arbitrary, including deployments multiple times a day and running different versions side by side while encouraging the adoption of the latest version

All that is great. You can now go from an idea to v1, to getting v2 to the customer in days. So, what is the magic of SRE? There is no magic; there is only a software development mindset applied to IT management.

Conservation of energy

I am very happy to admit I am the laziest person around when it comes doing the same thing twice. I want everyone else I work with to be lazy like that and I want everyone in the organizations that deliver services I use daily to be lazy as well.

I am happy to automate everything if I can. I've automated everything from deploying code, to answering email, to updating my manager of my weekly task completion at some point in my life.

> I am truly at the point in my life that once I learn that something is being done again, I must stop myself rolling my eyes, because this should have been automated. The waste of brain capacity is a terrible thing, and to the organization, it brings a headache because non-automated tasks scale linearly through the number of people needed, and the most expensive part of modern work is the people. If you or I are paying smart people, they should focus on new tasks that bring about agility and innovation, not repetitive work we should have already automated.

Simplistically(!), the SRE domain is about three things:

- Automate all the things that it makes sense to automate
- Everything to continue reliably working with changes happening all the time
- Ensuring total system observability, by using and testing it

That is it.

It is as true, for example, for testing and delivering code as it is for managing infrastructure.

For example, CI/CD must be used to deploy any software changes, to do so with minimal (if any) downtime, to do it reliably (without introducing issues and causing unexpected downtime), and to do it at scale across people and applications/services.

If we identify an issue in production that we've seen before, and we know how to resolve it, then there is no need to wake up anyone at 3:14 a.m. to look at it. The issue should resolve itself and automatically raise a ticket for someone to investigate the next day. Potentially, a permanent fix could be released in the next patch preventing the issue from happening, only if it makes sense to fix it. For example, when it perhaps doesn't make sense to fix it is if you are running third-party software that is on a slower cadence but will fix the issue, eventually. Or maybe it is better to update your documentation, runbooks, or best practice guides – it all depends what the issue is.

But you are still responsible for governing your SaaS platform, so you can't leave it in a failed state. Thus, automate the recovery and scale out the instances to cover the temporary loss of capacity.

A hugely welcome but unintended consequence of making the same people responsible for writing code, maintaining the platform, and getting woken up in the middle of the night when things go badly is you get better-quality code, you get automation, and you get fewer repeating issues.

I assume you don't like getting up at 3:14 a.m. and your SREs most certainly don't. So, if the code gets better, there is more automation and there are fewer issues – as if by magic. But again, there is no magic, only a software development mindset applied to IT management.

Let's look at what skill SREs have and how they apply them!? Usually, it's a valuable combination of a few or all of the following skills:

- Software development

- System administration

- Operational experience

- Security baked-in mindset

- Data analytics

You must already be thinking: where do I find these people and how can I afford to pay them? It is true that these are top-tier people, doing their very best in difficult and constantly changing circumstances.

The good news is you might already have people with these skills – some have one, some have two, and some may have all of them. You might also have people that are willing to learn continuously.

These are your starting team. If you don't have any, start looking for them. And yes, these people are paid well. But they are worth it. SREs stand between your company being left at the side of the road by your competitors. If your competitor has a reliable platform that is constantly and consistently evolving and you don't, well, then I've got news for you. We discussed earlier how cloud adoption requires a cultural and mindset change. Well, this is it.

SRE requires a mindset change and potentially a re-organization on your end. Gone are your silos; you now need a team that develops, deploys, and manages your services and does it quickly, reliably, and without waste.

A lot of you might think *We don't need this; we don't have such scale, or such agility required*. To that, we say perhaps you don't require such scale, but when it comes to agility, your competitors do have it, and if they don't, they will. What would you do then? Some of you will want to try it your way first, so come back here after you're done. We would welcome you back!

You will still have software developers, but your best ones (in our opinion) will be SREs – and they will all work closely together doing the following:

- Developing code

- Deploying services

- Monitoring services and the platform

- Ensuring availability and latency through any changes

- Responding to emergencies

- Improving and innovating on the platform, services, architecture and on ways of working together

If you are running a public-facing SaaS platform, there is usually a need for 24/7 support and/or management of the platform, so an SRE team is a great point to start to provide some governance to other teams on how to handover tasks from one time zone to another, how to document well, and how to pick and choose priorities to work on.

What the SRE team is there to do is to enable freedom and flexibility for the business and for the organization to innovate, build, manage, and deploy new and existing services across the world in a way that is agile but also well understood and patterned, so it reduces the overall stress levels across the organization.

SRE is also there to stop any deployments that would break SLAs – even if the downtime is a few seconds, deploying software updates hundreds of times a day may cumulatively be too much.

SREs are responsible for SLAs with customers, so they are there to ensure that quality code is delivered and minimal disruption is observed. SREs will always know the error budget available to different teams/services and will act accordingly, to allow more innovation or stop deployments altogether until the SLA is within the expected range.

This also greatly helps with other teams having more or less maturity in their processes, as it will be crucial to identify teams that are struggling and that need help.

SREs split their tasks between the following:

- Operations (keeping the platform running smoothly)
- Project work (features, scaling, and automation)

This is important because operations, while important, are a lot easier to do once you also know the inner workings of services. It also gives a reprieve from usually more stressful tasks in operations with (hopefully) less stressful tasks such as development.

To re-emphasize the difference between traditional IT and the SRE approach discussed here, consider the following.

Achieving 100% reliability is not the goal. The goal is to have as much reliability as needed, and almost always in enterprise organizations, the SLAs are too tight for no business reason – so look into confirming the SLAs. If an SLA is 99.99%, that will impact how often teams can deliver to production. If it is 95%, the teams may do a lot more innovation than with an SLA of 99.99%.

Consider that with an SLA of 99.99%, the monthly acceptable downtime is just over 4 minutes (this can be calculated here: `https://uptime.is/complex?sla=99.99`). With an SLA of 95%, it is 36 hours (`https://uptime.is/complex?sla=95`)!

These are just two *arbitrary* service availability SLAs to show you the difference between them. It is important to know these and the consequences in terms of acceptable downtime.

Extreme availability also comes at a huge cost. Other services dependent on such services will become evolutionarily complacent, meaning they will be less robust in case this service that never fails, fails.

This is not what you want, because that one time this extremely reliable service fails, you will have failures across the board because other services aren't robust enough. Kind of like people buying earthquake insurance after the earthquake.

Earthquakes?

You know how earthquakes work, right? Tectonic tension builds up over a long period of time and then it releases suddenly (a layman's explanation only – there is so much more to it but this isn't a book about earthquakes!). Well, people tend to buy insurance in the months and years after an earthquake, which doesn't really make sense. You should always have insurance if you can, but if you live in an earthquake-prone area, you should be more likely to buy insurance if there haven't been earthquakes in a long while – because this usually means one is (or more are) approaching. Buying insurance after the tension has released is either late or early – however you want to call it.

So, how does that tie in with the failure of an extremely reliable service? Well, over time, as the service is working perfectly, people and automations get used to it. They stop building insurance, building graceful degradation into their services – depending on a service too much to always be reliable, then when the service eventually fails (for whatever reason), you tend to discover that a lot of auxiliary services will fail in all sorts of catastrophic ways because they relied on the service too much.

One way to mitigate this is to bring these really reliable services down occasionally and deliberately to force these issues with auxiliary services to be identified and resolved faster and in a timed, planned, and at least semi-controlled environment. Otherwise, you will be building up the tension and when it releases, it will be like an earthquake – but for your entire service-driven architecture.

Because of all this, the ability to automate is the most important skill, so when hiring or educating your people to become SREs, remember that. You want people who will quickly become bored with repetitive tasks and have the skills to automate that boredom away.

Sometimes there is a fear that these folks will be unproductive through most of the day, which is both true and false. True in the sense that they won't be busy doing the same thing repeatedly, but false because they will now be more productive in other areas, such as recognizing and responding to issues before they happen, which is where their value really lies! There is a whole science of how "busy" each resource (including people and machines here) should be to be able to deliver the most value. Hint: it is not 100% busy – not even close. You must have slack built into the system or you will be unable to respond to changing conditions.

SREs are, by default, agents of rapid change and innovation, so they are one of your primary audiences when discussing and planning work on innovations.

SRE team members will focus on the following:

- Availability
- Performance
- Efficiency
- Monitoring
- Incident response
- Capacity planning

And if you look closely, where would these tasks lie in our modern world other than with SREs? IT was responsible for these previously, but now it is time to take the challenges of these responsibilities to a new level.

Because SLAs are usually tight, and humans bring latency to any response, the best response is always an automated one. Therefore, targeting the correct SLA is crucial.

Here are things to consider when planning an SLA:

- Is this a free or paid service?
- What are competitors offering?
- How critical is the service to customers' revenue?
- Is this service targeted at a consumer or an enterprise?

The cost of higher SLAs in terms of *more nines* of availability is exponential, so SLAs should be custom tailored to suit the market demands, just right.

You can always refine these over time as you get to know the system more and observe it in production.

To combat the issues when overachieving, you can also always take the system offline occasionally, throttle requests, or use some other mechanism to prevent overachieving!

Cloud governance

Now that the organization is working on the cloud, every day things are accumulating, for example, work tasks, application development, service migrations, performance metrics, documentation, and so on.

Are you in control? Is all this getting out of control? How do you know?

Are teams working with or against each other? At any point in time, do you know what everyone is working on, and is that the top priority at that point? Is there duplication? Is everyone happy with their role? Does everyone understand their role? Are teams and team members improving or stagnating?

Is delivery slowing down as complexity creeps in?

All this and more will be addressed here. You will learn to be in control while not micromanaging teams and tasks and empower teams to make decisions aligned with the mission, vision, and strategy.

Assess your current state and plan the evolution of governance while the evolution of your platform is ongoing. Expand into more services, more clouds, and more vendors and still stay in control. How can you ensure existing governance is always respected?

It's likely you will need to approach this very differently from existing management/control practices, and this is crucial to the adoption success, and deserves special focus and planning.

There is a crucial difference between the overall general governance of/in your organization and Azure governance. Let's discuss both!

Overall organization governance is all about staying on top of changes, drift, or – as you might want to call it – *entropy*. Entropy is often interpreted as the degree of disorder or randomness in a system, a lack of order and predictability, or a gradual decline into disorder.

As you might imagine, a system first must be in some sort of order before it can gradually decline into disorder. Here, we will take for granted that you've gone through the processes of cloud adoption, have made some sense of it in/for your organization and for the roles in your organization. The issue than becomes how do you keep that order? People will come and go; projects and services will start up, peak, and reach their end of life. How do you ensure as things change that gradual decline into disorder is prevented?

The bad news is that entropy is inevitable in our universe (and likely in all other universes as well), and your little organizational bubble is no exception. The good news is that the system is rather large (on the scale of the universe, or even on the scale of Earth or your organization's market, or even your organization itself). Locally, you can fight entropy at least for a while.

Key to a successful organization governance model are the following (in order):

1. A model accepted by everyone, understood by everyone, and a model that works for the business.

2. Agility to bring that model into practice at every interaction – in engineering, in product management, and in architecture. But, as always, and most importantly, for the business. If the business can change direction/priorities and it takes you at most a few days to course-correct (or a few weeks for the craziest of changes), then you are agile. Your governance model doesn't necessarily also have to change, but which ever model you have must fit your business and deliver results quickly.

3. A documented governance model that ensures everyone is still aligned, and everyone understands it and their role in it.

4. A scalable governance model that enables different parts of the organization to work together, without annoying, burdening, or slowing them down because there is governance in place.

Now, 1 is hard, 2 is equally as hard, but 3 and 4 are tricky. For the time being, you can reach a conclusion for 1 and 2. They will need to slowly evolve over time, but you will reach a logical conclusion, which will mark their completion. With 3 and 4, your work will never be done. Finding the right amount of governance, like Goldilocks looking for the right bed (of *Goldilocks and the Three Bears* fame) is very tricky.

There can be no compromise of the best practices your organization has set, across all five pillars of the Azure Well-Architected Framework:

- Reliability
- Security
- Cost optimization
- Operational excellence
- Performance efficiency

These are all easy. You just need to schedule, attend, and act upon the results from regularly scheduled reviews. It's desirable for these to be a part of all teams' schedules.

For example, if you do a regular top-level review of budgets (in Azure terms) and costs, you can identify surprises early on. If each team does this review for their services regularly, they can not only identify the cost optimization items they can improve upon themselves but also report them to the top level, which then ensures that you can have a light touch on governance. You might not need a well-regulated and bi-weekly scheduled cost optimization session. You can have much fewer formal discussions around the watercooler and hence, your governance works and it isn't overbearing.

Encouraging all teams to do this is great, and once they realize the alternative is a lot more meetings, discussions, and feedback from you, they will take regular reviews into their team meeting cadence. For example, the cost might be regularly discussed by finance (at a service level), by your partners, by your legal team, and so on.

The cost of duplication

I've worked on many projects where the impact of cost saving came from other teams. The product management team may have oversold the availability requirements of that document generation service, or the legal team may have identified a change in the law that allows you to store backups in a cheaper jurisdiction.

One area where this is especially useful is any sort of migration project built on top of the Cloud Adoption Framework. I've seen plans go from moving three data centers worth of VMs into the cloud, only to realize that about 10% could be turned off outright, 10-20% could be moved to lower-spec machines, and about 5-25% of services across such an estate could be consolidated. Do you really need nine different document generation services?

This is not to be underestimated. We can go from moving 200,000 VMs 3 times to 150,000 VMs 3 times. That saves a whole bunch of effort, both in person-hours and direct costs of infrastructure.

And more than once I've seen actual "misalignment" with existing laws that had to then also be corrected during a migration project.

Another important lesson from this is that governance itself is the responsibility of everyone. If your organization is happy to rush to architects to cost-optimize, they need to be set straight that everyone is responsible. The same goes for the other pillars as well.

The methodology of the Azure Cloud Adoption Framework centers around the concept of a landing zone. A landing zone provides workloads with a common set of services that every workload will need to use (to integrate and/or communicate with other services and/or share the infrastructure).

An Azure landing zone is a setup of Azure subscriptions that serve as the foundation for all other workloads. You cannot deploy anything into Azure without a subscription. Aside from being tied to payment methods, they provide a segregated space for workloads. As you can have as many of these subscriptions as you like, it might become difficult to manage them once they proliferate through the organization. Knowing which subscriptions have which workloads, who is paying for them and how, as well as who is responsible for each of them should be a part of your governance plan.

As the landing zone needs to be scalable and modular, there may be a place in these subscriptions just for these common services. This is also a great initial responsibility for your SRE team, which you are starting to build.

To be able to define your governance model, you need to define your organizational policies to cover the following:

- Business risks
- Regulatory compliance
- Processes
- People

These are all important to understand and plan for.

How will you evaluate and monitor costs and create accountability? How will you establish security red lines and base lines? How will the creation of resources be done in a consistent manner? How will identity management be taken care of (internally and externally)? How will you accelerate the amount of services and workloads you have and their deployment? These are all questions you must answer.

The bad news is that getting this wrong now will incur a lot of technical debt, which may or may not be easy to fix later (and Murphy's law says it won't be). The good news is that while this might seem daunting, it really isn't.

Once you start planning for these things, you will see you only need a few things in Azure for the landing zone to be ready. It can always evolve, but it can start small.

If you are a part of an enterprise organization that is moving into Azure, Microsoft and their partners can help you do this quickly and easily – if they have cooperation from your organization.

If you are not a part of a large organization, you can do this quickly yourself. The hard parts are always the following:

- DNS
- Networking
- Security

If you are a smaller organization, your DNS will usually be quite simple (a few domains and maybe a few subdomains), networking won't be involved to connect hundreds of offices, data centers, and different clouds, but security is always hard (no need to sugar-coat it here). Also, DNS and networking will eventually be *complete*; security is forever on your plate.

The easiest way to begin is with the end. How do you want or how do you need the end state to look in management or project management terms? You then look into listing the gaps and how to close them. This might be an involved process for some and might be very easy for others. For example, if you are a start-up with a new idea/product on the market, your landing zone may be tiny when you start but may expand over time. It might only have things such as the following:

- Azure Kubernetes Service with Azure Storage and Azure Cosmos DB
- Azure Service Bus
- Azure DDoS Protection
- Azure DNS
- Azure Virtual Network with Network Security Groups
- Azure Traffic Manager

Setting up these common services is a magnitude easier than adopting the cloud on an enterprise scale, but it is not better or worse; it is exactly what you need: no more, no less.

There are a few things to understand as they are the anti-patterns of this approach. In the cloud, a lot of responsibilities are shared – for example, security or service updates or backups. Yes, the cloud can do a lot for you, but don't forget to plan for these yourself – and test them regularly.

In Azure, luckily, you have the controls that you need. Let's go through some of them:

- Azure Privileged Identity Management
- Azure management groups and Azure policies
- Azure Resource Graph and Microsoft Cost Management

Azure Privileged Identity Management is a service that enables just-in-time roles to be activated (confirmed with an MFA prompt), which means that no one has default access to any resources in Azure. To get access, activation (and maybe even confirmation by another) is required. An MFA prompt must be completed and only then is the access given to specific resources, roles, and permissions for a maximum duration determined by you. Here is a tip: start using this service today!

> **So many roles...**
>
> I have about 30 of these roles now that I am eligible for. You too can have as many as required, and you can activate them as needed. This design also conforms to the least-privilege principle – you use the role that gives you the minimum privileges required to do the job you need. Usually, role activations are needed for investigations or education sessions and hence are read-only. Everything else that involves making changes is done through CI/CD pipelines. You are using CI/CD, right? For everything, right?

Azure management groups allow you to combine subscriptions into groups that you can then apply the same policies on. For example, development and production groups versus a sandbox group. You can, I hope, imagine why different policies would be needed for these?

Azure policies allow you to specify conditions that will be evaluated before any changes to environments are made – for example, in development subscriptions, you can limit the number and tier of services allowed to be deployed, which will be different for production. In development, you will use lower-tiered services (except in some tests), while in production, you want only the enterprise-grade services. In production, you might also want to limit the regions you can deploy to if, for example, there is a legal requirement for customer data to never leave a particular jurisdiction.

Policies are evaluated before deployment is allowed, so it saves you and your governance plan from including sections on checking and double-checking deployments – policies will prevent non-compliant deployment from ever occurring, provided you take the time to set up the policies.

You can even have alerts set up and attempted policy violations can be reported to you. This gives you another chance to govern only when needed. Perhaps it was a misconfiguration or perhaps it was a misunderstanding on behalf of the team member running the CI/CD pipelines, so talk to them and set them straight if needed.

Some of these services are cloud-agnostic so can be used across clouds in case you happen to be using AWS and GCP as well as Azure.

Communication planning

How can you bring the entire organization along for the ride?

Some things will go well and some won't. Either way, it is important to have a plan for communicating the good and the bad news, both for the same reason – so everyone can learn. This is the human element of transformation and adoption.

A communication plan is a plan that goes into minute detail on who communicates what, when, and of course, why.

Which communiques will need to be sent, by whom, and at what cadence. Also, why? Because the content must match the target audience, must be timely, and must be useful.

Err on the side of overcommunication (even if it's just initially) as people will be confused, unsure, and timid or will just need what they think is true reaffirmed in their minds. Keeping everyone on the same page, even if just by Ccing everyone in the communication, is important.

Orchestration and synchronization are important, to a point. What you are aiming for is for every team to know what they need to know to execute the plan. And it needs to be clear, specific, and, above all, unambiguous.

The point is less that orchestration and synchronization are important as a process and more that once the message is understood, no further orchestration and synchronization is needed as everyone can get on with their tasks. Touch points to verify that everyone is on the right track should be much lighter than a full-on orchestration and synchronization approach that takes a lot of time away from delivering on the business goal.

For example, once the teams are aware of the platform target architecture, unless it changes, there is little benefit of having all teams on architecture boards twice a week. They got it the first few times you showed it to them and they can now work on their independent services by themselves. If teams understand how their APIs and API versions should look, which service will host them, and how the APIs are scaled and secured, they can develop as many services to that standard as the business requires. That is the agility we are all chasing.

So, when should these communication activities happen?

- When planning, then when launching new service.
- When starting a new research project.
- When initiating any sort of large(ish)-scale company changes (re-orgs, as they're known to some).
- When changing existing processes (hopefully, to increase agility and ease of understanding and use, rather than introducing more delays).

- For customer and market messaging! Your organization is changing, and none of the changes can be hidden from customers. For example, they used to get a perpetual license and a new version of software every quarter. Now, they pay as they go and get new versions of different services multiple times a day.

Communication goals and objectives must be clear and they must be communicated in a way that the target audience is willing to consume them. PowerPoint? An API documentation portal? Video? Twitter ads? Or, multiple channels? Whichever medium works best.

Review how the message is received and re-evaluate the communication plan going forward.

To do all that, you must navigate your organization structure (and sometimes, you need to reorganize it). Alerting and triage procedures should be devised and they should be very light. They must be designed to alert you of gaps.

For example, if all teams are delivering new service updates daily or weekly, but some teams are doing it monthly or quarterly – why!? It might make perfect sense for those services, but if you don't know it, an alert should be ready to remind you. Then, you and your organization can flex interpersonal skills and check up on the deviation.

The processes must be simple enough that everyone can follow them, including new joiners, the legal team, SREs, and the CEO.

The tone of communication is also important. Different tones for different audiences, usually a professional tone but with a whimsical touch, is best to both convey the information and not feel like a demand.

Communicating during an emergency

As for the communication plans for emergencies or incidents, you must establish clear boundaries of responsibility. If, during an incident, a team member is tasked with a task that is not explaining the CEO what is going on, then they need to be empowered to say *Can you please check that information with our incident communication officer.*

If a team member is feeling cheeky, they can also offer to forward the communication plan to the CEO if they haven't had the time to check it out. It is only a few slides, and the CEO should be up to speed in 5-10 minutes max. The reasons behind this are numerous. Let's start from the beginning though. It is important to know when to communicate and when not to communicate.

While being proactive is usually great and should be encouraged, in the case of platform events it is strongly discouraged. There are multiple reasons:

- Your customer may not be aware or care (for example, outside of working hours).

- Your customer may not notice (depending on the event – full-scale event, performance degradation, or a partial event).

- Your customer may not be impacted (due to the high level of automation, segregation, etc.). An event may be high priority for some customers, and for some non-existent.

- Your customer has not reached out (for whatever reason).

- SLA breaches are usually counted on a case-by-case basis and from the time of the report (except in exceptional circumstance when we approve an SLA breach for all customers).

- The customer may not be aware or is not impacted, so there's no reason to panic them.

- An outage may have a wider impact – for example, an internet provider or internet-wide outage. So, our event might be a minor one compared to the greater issue and may not be noticed or acknowledged.

Alright, so now the question is *when to communicate*.

There is only one reason to communicate with your customer directly about an event (it goes without saying that if you have a health or status portal, you should update that as per your communication strategy) – the customer has reached out (phone, email, IM, etc.) or has opened a ticket!

There is one other scenario *when* you should (must!) communicate with your customer and that is *before* any event! Why?

- The customer must know what the procedure for SLA breach reporting is before any such event happens

- The customer also must know where to go for information during an event, and what to expect as communication during an event (which is very little; as we will be working on mitigating the event, not on communicating with individual customers)

- Outside the information the operations team provides, they should expect nothing more until the event is resolved

- After the event is resolved, they can and will get all the information we can provide (excluding sensitive customer data, it should be a complete picture of the event)

- After root cause analysis, they can and will get a detailed report as to what steps we are taking so that the event doesn't reoccur

You *must* not go into any further detail other than what the official communication is, even though you might want to because you might want to either preserve the relationship, to help them out, or just to be comforting during a stressful time. You *must* resist all that. Why? You may inadvertently lead them down the mitigation path that may do more harm than good, either due to poor communication during a stressful time, due to their poor understanding of what you are communicating, or due to preliminary information being wrong, just to note a few reasons.

You may expose the company to excessive damages and/or other liabilities, both directly to the customer or to regulators or governments, especially if there is a data breach. The goal is not to hide information but to report exact, truthful, and detailed information as and when it is available – during an event is not the time.

When you communicate during an event but cannot also communicate a solution or timelines at the same time, it leaves the customer helpless and powerless, and while there will be an impact on your reputation (however large or small), any back and forth (especially as we learn more and correct ourselves in the process) will only serve to undermine your capabilities with no benefit for you or the customer. You cannot communicate any more as you don't know, or don't know with certainty, that what you are communicating is truthful and helpful.

A final thought: working with people is hard, but there are patterns (some very counter-intuitive) that can help you learn about how to communicate and communicate better with different stakeholders and different audiences. These steps and templates allow the process of communication to flow smoothly and, above all, consistently and transparently. Since only you can prevent forest fires, only you can know your organizational culture to be able to ebb and flow through its hierarchy effectively and deliver news – good, bad, or neutral – efficiently.

Summary

Organizational transformation is essential for any organization that wants to remain competitive and successful. It involves changing processes, systems, and policies to stay up to date with new trends, technologies, and customer needs.

Site Reliability Engineering (**SRE**) is a discipline that combines software engineering and operations to ensure the availability, performance, scalability, and security of an organization's computing systems. By adopting SRE best practices, organizations can achieve better application reliability with fewer resources while providing a more resilient platform to support their business goals. Additionally, they will be able to respond faster to any incidents or outages that may occur to reduce downtime and mitigate potential risks.

Organizations need to develop communication strategies that are tailored to the event or situation at hand. Communication should always be done before an event so customers know what to expect and where to go for information. After an event is resolved, customers should also be provided with a detailed report on the root cause analysis and the steps being taken to prevent the same incident from happening again. It is important to resist the temptation to give any more detail than necessary during an event, as this could lead to confusion, incorrect expectations, and further damage. By following these steps, organizations can ensure that their communication is consistent, transparent, and tailored to the situation. This will help foster relationships with customers while also ensuring that the company's reputation is maintained.

Part 3:
The Execution and Iteration

After all of our careful planning, the realization of these efforts is now tangible. Everything that came before this was trial-and-error learning experiences. We are actively embracing cloud adoption and handling workloads; however, it's essential to ask ourselves whether we're merely maintaining a status quo or striving for optimal performance. In order to guarantee consistent scrutiny while still allowing room for improvement, how can we accurately measure progress as well as provide an impetus for further advancement? These topics will be addressed in detail throughout this part of the book.

This part of the book comprises the following chapters:

- *Chapter 7, Creating a New Operating Model*
- *Chapter 8, Architecting Your Cloud*
- *Chapter 9, Closing Thoughts*

7

Creating a New
Operating Model

The cloud has revolutionized the way business IT operates, making it easier than ever to access and deploy hyperscale technology. But in order to get the most out of your cloud investments, you need an effective operating model that takes advantage of the agility a cloud environment can offer. The Azure Cloud Adoption Framework provides guidance on how to create and implement a successful cloud operating model that will help you optimize performance and maximize value from your cloud initiatives. By following this framework, organizations can ensure they are taking full advantage of their resources while avoiding common pitfalls associated with cloud adoption.

In this chapter, we are going to cover the following topics:

- Cloud operating models
- Landing zones
- Migration

Understanding cloud operating models

A cloud operating model is a framework that helps organizations maximize the value of their technology investments. It is designed to provide an overarching view of how the organization uses, manages, and secures its cloud resources. The operating model defines roles, responsibilities, processes, policies, and tools necessary to manage technology operations effectively and efficiently.

At its core, a cloud operating model provides guidance on how people and technology should work together to achieve the desired business outcomes. An effective cloud operating model also allows organizations to proactively identify any potential issues such as security risks or compliance violations before they occur. Another focus area of a cloud operating model should be to maintain efficient operations through effective management of costs and complexity. This can be achieved by planning and designing the cloud setup early, standardizing how services are configured, and revising existing IT operational processes, eliminating redundant controls and procedures that limit agility.

In order to ensure success with their cloud operating model, organizations must define and implement several key elements, such as roles and responsibilities, governance and compliance, service delivery management, and training and education programs. Each element must be carefully implemented for it to be effective in delivering value for the organization. By following a well-defined cloud operating model framework, organizations can ensure that they are mitigating risks and staying compliant with industry regulations. Over the next few sections, we will dive deeper into each of the key elements of a cloud operating model.

Roles and responsibilities

Cloud operation roles and responsibilities are essential to ensure success when managing cloud services. A well-defined cloud operating model should include roles and responsibilities for each of the key stakeholders involved in the organization's technology operations. These roles should be clearly outlined, with responsibilities assigned to each role and individuals accountable for their actions.

The most critical role in cloud operations is that of the cloud operations manager, who is responsible for setting up the operating model, managing its implementation, and ensuring it remains effective over time. This person should have a deep understanding of cloud technologies and be able to coordinate between teams to ensure smooth operations. They must also have excellent communication skills to effectively manage stakeholders and collaborate with (but really manage) vendors.

Other key roles within cloud operations include the following:

- *Cloud security officer*, who is responsible for protecting the organization's data by ensuring security best practices are followed
- *Cloud architect*, who designs the overall structure of the cloud infrastructure
- *Cloud administrator*, who maintains the day-to-day operation of applications, services, and solutions
- *Cloud platform engineer (DevOps)*, who bridges software development with operations by setting up automation to construct platform components such as virtual machines, databases, and networks

Each of these roles has its own set of duties that must be carried out for an organization's cloud operations to run smoothly. By having a clear understanding of these roles and responsibilities from the start, organizations can avoid in-fighting, conflict, and wasted effort.

Governance and compliance

Governance and compliance is an important part of any cloud operating model. It ensures that the organization's technology operations are secure, compliant with industry regulations, and adhere to best practices. Cloud governance and compliance involves creating policies and procedures around data security, privacy, access control, managing cloud resources, and more. It also requires regular monitoring and reporting to ensure that the organization is meeting its objectives.

Cloud governance starts with establishing a set of standards for how cloud services should be accessed and used within the organization. These standards should include guidelines on topics such as roles and responsibilities; separation of duties; access controls; data protection; identity and authentication requirements; encryption techniques; logging technologies; patching processes; backup strategies; disaster recovery plans; regulatory compliance requirements; and more. Once these standards have been established, they must be implemented across the entire organization using a wide range of tools, including automation, policy enforcement systems, identity management services, auditing software, and more.

Compliance is another key aspect of a cloud operating model. Organizations must be able to prove their adherence to industry regulations or risk penalties such as fines or the loss of business. To meet this requirement, organizations must have stringent processes in place for tracking changes made to their cloud environment over time, as well as continuous monitoring to detect potential violations before they occur. This includes implementing automated policy checks at regular intervals to ensure that all systems remain compliant with all applicable laws and regulations as well as ensuring that data always remains secure.

Moreover, organizations should assess their current cloud infrastructure on a regular basis to ensure system integrity and minimize risk from potential vulnerabilities or malicious actors. They should also regularly review their policy documentation to ensure that it is up to date with industry best practices while also conducting audits to check whether policies are being enforced correctly across the board. Finally, organizations must remain vigilant against any new regulations or legal updates that could impact how they manage their cloud operations, to stay ahead of any potential compliance issues that may arise in the future.

Service delivery management

Cloud service delivery management (or operations) is the process of managing and maintaining cloud infrastructure to ensure that cloud-based services are delivered in a timely and efficient manner, while meeting all business requirements. It is a critical component of any successful cloud operating model and involves several activities, such as planning, designing, provisioning, monitoring, and updating cloud resources to ensure optimal performance, security, and cost efficiency. It is essential for any organization that relies on cloud technologies for business-critical services or applications.

Delivery management starts with the planning process. During this phase, important decisions need to be made about which services will be used by the organization, who will be responsible for delivering those services, and how many people are required (resourcing). Once these decisions have been made, the organization can move on to the design phase, which involves creating an architecture blueprint that outlines how each service should operate within the environment, including components such as storage, networking, and compute resources.

The deployment of all necessary services should be configured so that each can run optimally within the environment. This includes setting up access control policies, configuring security groups, and other administrative tasks. It is important to test services to ensure that they function correctly before releasing for general use. Automation is vitally important for streamlining and simplifying many of the more time-consuming tasks associated with managing a cloud environment, including resource provisioning, configuration changes, logging, and alerting for potential threats or vulnerabilities, as well as automated testing and compliance checks. Automation can also eliminate the risk of errors caused by human intervention during configuration tasks or deployments.

Many traditional IT operations best practices still apply to the cloud – do not throw them out just yet. For example, comprehensive change management processes that allow for risk assessment prior to making changes, and recording changes made over time, help identify the root cause of issues when they occur. Of course this can sometimes result in tunnel vision, focusing just on a recent IT change as the likely cause of an issue, having good monitoring and health checks are invaluable when things go wrong. Regular policy checks/audits are important to avoid drifting too far away from the baseline design, but rather than relying on *checkbox exercises*, the cloud enables automated policy checks to be run frequently or whenever a change occurs.

Once everything is up and running, monitoring and maintenance needs to be performed regularly to keep cloud services running smoothly over time. Why do we need maintenance? Surely, the cloud provider takes care of this, no? Well, yes and no. To avoid customer disruption, many services are versioned, and customers must plan when they will upgrade their deployment. It is common for cloud providers to offer new virtual machine options that perform better or are more cost effective than what was previously available. It is important that you keep up to date with the cloud as it evolves. Also, virtual machines, just like their physical counterparts, require regular maintenance to ensure the OS and applications are kept up to date with the latest patches for critical vulnerabilities and maintain integration with evolving cloud monitoring tools.

Overall, cloud service delivery management helps organizations run high-quality services reliably by minimizing potential downtime or disruption caused by technical issues or changing business needs. By having a well-defined process in place and leveraging automation, organizations can ensure smooth operations over time while also reducing costs associated with manual processes or errors caused by lack of oversight.

Keeping operational documentation updated

It is essential to first create operational documentation in order to ensure efficient and effective delivery of cloud services. Types of documentation include process documents, system diagrams, system configuration guidelines, procedures for accessing and using the cloud, security policies and procedures, user guides, and training materials. Process documents should contain detailed information on how the system works and which processes should be followed to ensure smooth operation. System diagrams provide a visual of the overall architecture and can be useful for troubleshooting purposes. System configuration guidelines are necessary to ensure that all components of the cloud environment are configured properly and securely, while procedures for accessing and using the cloud will help users understand how to properly use the services available in their environment. Security policies and procedures must also be documented to ensure that all access is controlled in a secure manner.

Once the documentation is created, it needs to be regularly maintained and updated as new features are added or existing processes change. This ensures that everyone in the organization has access to accurate and current information about the cloud infrastructure, enables better collaboration among team members, and potentially improves decision-making. Well-managed documentation also provides an up-to-date record of changes made to the cloud environment, which can be useful for troubleshooting issues that may arise. Of course, for most environments, it is important that you have evidence to support compliance with regulatory requirements and provide a clear audit trail of activities related to the cloud. With proper documentation in place, you can ensure that the entire team is on the same page when it comes to using the cloud and can collaborate effectively to make sure all operations are running optimally.

Reviewing training and education programs

An effective cloud training and education program is essential for any organization that wishes to maximize the benefits of the cloud. Training should be tailored to the specific role within the cloud operating model and appropriate to the level of technical ability. A comprehensive training program should include both classroom-style lectures as well as hands-on, real-world exercises to ensure that your people understand the nuances of working with the cloud.

For those who are new to cloud computing, initial training should focus on understanding key concepts such as the basic cloud fundamentals, compute virtualization, cloud infrastructure versus traditional IT infrastructure, and cloud automation (for example, infrastructure as code). Lectures should cover topics such as cloud architecture, scalability, cost management, security, and deployment strategies. Hands-on exercises should involve setting up a basic cloud environment with web serving or managed databases and learning how to configure services such as load balancing, auto-scaling, backups, and network access policies. Additionally, students should be taught how to troubleshoot common issues experienced with core cloud services.

For those in development or operations roles, additional topics can be discussed such as DevOps/SRE practices, cloud management APIs, and scripting languages used for automation. In addition to lectures and hands-on exercises, more advanced users can benefit from lab environments that allow them to explore various tools and technologies in a safe sandbox environment before migrating workloads or touching production systems. This will help them develop an understanding of how different components interact with each other, allowing them to effectively utilize different services within the cloud ecosystem.

For an effective cloud training and education program to be successful, it is important for organizations to invest in high-quality materials that are constantly updated. This includes books, online tutorials and video lectures from industry experts who have experience working in enterprise-level systems.

Additionally, organizations must ensure that they have enough resources available within their teams so that everyone has the time to learn and access to materials they need.

The Microsoft **Enterprise Skills Initiative** (**ESI**) is an effort to digitally empower people by providing them with the tools and resources needed to succeed in the Azure cloud (and many other Microsoft products). The ESI provides tailored training programs, educational materials, and best practices for enterprises of all sizes. The program emphasizes the importance of developing an understanding of cloud computing fundamentals as well as real-world implementation scenarios for the Azure cloud.

The goal of the ESI is to help organizations increase their skills, agility, and improve knowledge across a variety of use cases, such as cost management, application modernization, DevOps, and machine learning. To achieve this, the program offers a comprehensive suite of resources organized into three key areas: Learn, Virtual Training Days, and Microsoft-Delivered Courses:

- *Learn*: Provides access to several learning paths for different roles. The materials provided support self-learning through short, step-by-step tutorials and interactive scripting environments.

- *Virtual Training Days*: Provide a variety of one-day online lectures and lab environments for hands-on exercises covering everything from Azure fundamentals, architecture considerations, cost management, and cloud migration to advanced topics such as using Azure services for big data and analytics.

- *Microsoft-Delivered Courses*: Provide numerous multi-day classroom-style training led by expert Microsoft instructors. These courses offer guidance on best practices such as how to design highly available architectures with built-in redundancy, how to leverage serverless computing or containerization strategies, and how to build secure solutions and approach data privacy regulations such as the GDPR or HIPAA.

GDP what?

The GDPR and HIPAA are important because they help protect sensitive customer data. The GDPR, or General Data Protection Regulation, is a regulation in the EU that governs how companies handle personal data of individuals within the EU. It applies to any organization that processes or stores the personal data of EU citizens, regardless of where the organization is located. HIPAA, or Health Insurance Portability and Accountability Act, is a US law that sets standards for protecting sensitive patient health information. It applies to healthcare providers and their partners who handle this information.

Both regulations require organizations to implement strict security measures when handling personal data, such as encryption and access controls. Failure to comply with these regulations can result in significant financial penalties and reputational damage. In the cloud, businesses must ensure that their cloud service provider also complies with these regulations, as they will be responsible for protecting the data stored on their infrastructure. This means carefully selecting a provider that has strong security measures in place and regularly audits their systems for vulnerabilities. But don't fret, Microsoft Azure has you covered. Visit the Microsoft Service Trust Portal to find out more: `https://servicetrust.microsoft.com`.

Overall, the ESI program aims to empower enterprises so they can make informed decisions around their technology investments and successfully adopt and leverage the cloud, while also increasing employee skills and productivity. Training materials are tailored to every role that has a stake in the cloud and materials are updated regularly to match Azure technological advancements. As such, the ESI is a valuable tool for organizations of all sizes to support their cloud adoption strategies. Next, let's look at how structuring your cloud can help enable smooth governance and operations.

Understanding landing zones

A landing zone is your overarching cloud architecture that incorporates multi-subscriptions, identity management, management groups, policy and compliance, security, networking, high availability, and automation to allow your organization to quickly and securely deploy cloud workloads in Microsoft Azure. Typically a landing zone consists of core components, including Azure Active Directory, Azure Policy, and Microsoft Defender for Cloud.

What is an Azure subscription?

An Azure subscription is a distinct area for a grouping of cloud services and resources. Subscriptions are available in different tiers depending on the needs of the user. The Free account provides users with a low-cost entry point to test out Azure with 12 months of access to limited services at no cost. There are also multiple pay-as-you-go plans available for businesses who require more comprehensive services. The Enterprise subscriptions are designed for larger organizations that require more tailored agreements either as existing Microsoft customers or for industry-specific regulatory compliance needs, for example, audit rights.

When designing your subscriptions, it is often wise for cloud adoption projects to allocate sandbox subscriptions to provide a safe environment for testing and development. A sandbox offers an isolated environment from production environments where users can safely test out new ideas, configurations, and changes without any risk of disruption. A sandbox does not differ from any other subscription in terms of access to all the necessary resources, such as compute, storage, and networking. This helps developers ensure that their applications are running smoothly before pushing them out into production. However, cloud administrators must still control access, monitor usage, and remain vigilant in protecting against malicious activities. This can be challenging within an ever-changing environment, but the agility gains make it worthwhile.

Azure Active Directory

Azure Active Directory (AAD) is a cloud-based identity and access management solution offered by Microsoft as part of the Azure suite. It is designed to provide secure authentication, authorization, and compliance services for cloud, web, mobile, and desktop applications in an enterprise environment. AAD provides organizations with tools to create custom policies that can be used to control user access to resources while restricting undesirable activities.

With AAD, users can access multiple applications from any device or platform using a single sign-on experience. This makes it easier for organizations to manage user accounts and access rights across their entire range of applications. Additionally, AAD provides users with multi-factor authentication capabilities, which help improve security by requiring additional verification such as biometrics or one-time codes for logging into systems.

Additionally, AAD offers advanced identity protection features that help protect against malicious activities such as unauthorized logins or account takeover attempts. These features include multi-factor authentication, risk-based assessments that detect suspicious activity, adaptive authentication policies that scale with the level of risk associated with each login attempt, and automated remediation actions that can take immediate corrective measures when required.

AAD also offers comprehensive audit logging capabilities that enable administrators to monitor user activities within the system and identify any potential security threats in real time. Furthermore, AAD integrates seamlessly with other Microsoft services, allowing organizations to quickly set up single sign-on experiences across Office 365, SharePoint, or Dynamics CRM. Finally, it supports integration with a wide range of third-party identity providers such as Google Auth or Okta, allowing users to easily authenticate using existing credentials from other popular service providers. Organizations can ensure secure authentication experiences for their users no matter what platform they're on.

Examining Azure management groups

An Azure management group is a cloud organization tool that allows administrators to organize and manage multiple Azure subscriptions within a single hierarchical structure. It provides administrators with a centralized view of their resources, allowing for better governance and control over them. By using Azure management groups, organizations can easily use policies, access control, and compliance settings across their entire cloud, but also tailor rules for different workloads within a management group.

Management groups allows you to group multiple subscriptions together, while still maintaining the ability to drill down into individual resources. This means that resource users have an improved understanding of the cost associated with each subscription or resource within the group, as well as being able to identify potential threats or anomalies across the entire organizational structure. Additionally, this feature enables administrators to apply access control policies such as **role-based access control** (**RBAC**) across all resources in one go. This helps ensure that only authorized personnel have access to the appropriate resources, providing organizations with an added layer of security for their cloud environment.

Organizations can also use Azure management groups for cost optimization and budgeting purposes by setting spending limits on specific groups or even individual subscriptions within the hierarchy structure. These limits can be used to track actual versus forecasted costs over time to help organizations make informed decisions when it comes to cloud expenses. Furthermore, management groups offer support for conditional access policies, which helps ensure that users are accessing data from approved sources only when signing in from untrusted networks or devices.

Considering Azure Policy

Azure Policy is a service offered by Microsoft that allows you to define and enforce organizational standards and operational procedures across cloud resources. This platform provides access to an extensive library of built-in policy definitions as well as the ability to create custom policies. These policies can then be used to ensure compliance with regulations, secure data, and optimize costs while still meeting organizational objectives.

Azure Policy allows you to set restrictions on aspects such as resource types, locations, tags, and certain configurations for cloud resources. It also offers the ability to control user-defined access roles for specific accounts or groups using RBAC. This ensures only authorized personnel can make changes within their Azure environments. Additionally, it assists in preventing unexpected cost increases due to unused or unnecessary resources by providing visibility into usage metrics and allowing you to set limits on consumption.

Azure Policy also features automated governance capabilities, which enable users to quickly deploy pre-configured policies that guarantee compliance with industry standards and government regulations such as ISO27001, HIPAA, or the GDPR. Furthermore, it includes tools for continuous assessment and alerting that allow admins to act if any existing policy violations occur within their environment. Azure Policy is a powerful tool that enables organizations of all sizes to maintain a secure and compliant IT infrastructure in line with their organizational goals while still taking advantage of the many benefits offered by the cloud.

Protecting your cloud with Microsoft Defender

Microsoft Defender for Cloud is a cloud-native security service that provides monitoring and protection for an organization's cloud resources. It helps to identify threats, detect malicious activities, and respond to incidents quickly. This service leverages AI-driven analytics to continuously monitor an organization's Azure environment and alert them when suspicious activity is detected.

Microsoft Defender for Cloud provides both monitoring and preventative controls by leveraging real-time threat intelligence from Microsoft to facilitate the early detection of potential threats. It also allows you to customize detection capabilities through the use of custom rules that allow the definition of exactly what types of resources and activity should be monitored. Furthermore, it offers automated response capabilities that allow organizations to set up automated responses in the event of a security incident or breach, such as disabling accounts or restricting network access.

Bringing it all together

Azure landing zones are a best practice for organizations that are looking to quickly and efficiently deploy cloud resources in a secure, compliant, and cost-effective manner. A landing zone is an environment that contains all the necessary components for deploying cloud applications and services such as virtual machines, networks, storage accounts, databases, and so on. It also provides foundational guardrails that protect against misconfigurations and ensure the organization's cloud environment remains secure and compliant.

When designing an Azure landing zone, it is important to consider the organization's specific goals, requirements, compliance considerations, security policies, and processes. This will determine what type of architecture should be implemented as well as which tools should be used. One recommended best practice is to review reference architectures provided by the **Cloud Adoption Framework** (**CAF**), partners, and existing clients.

Once a landing zone has been implemented, it is important to review all of the resources on a regular basis. This will help ensure any misconfigurations or security vulnerabilities are identified quickly so they can be rectified before they become a problem. Additionally, it is beneficial to use DevOps practices such as automation, continuous integration/deployment pipelines, and infrastructure as code to reduce the manual effort associated with managing changes in the cloud environment.

Organizations may also want to consider leveraging other security services offered, such as Microsoft Defender for Cloud, which provides layers of protection against potential threats or malicious activities targeting their cloud infrastructure. Additionally, you can make use of the Azure Security Center, which provides centralized visibility into the overall security posture across multiple subscriptions, along with recommendations on how to further strengthen it.

Azure landing zones provide an easy way to quickly deploy cloud resources in a secure and cost-effective manner while still maintaining compliance with industry standards or government regulations such as ISO, the GDPR, or HIPAA. By building best practices into the design, regularly reviewing resource configurations, utilizing DevOps practices for managing changes in the environment, and leveraging additional services such as Azure Policy and Microsoft Defender for Cloud, you can scale your cloud adoption while ensuring compliance and that data remains protected from potential threats.

Next, we will consider how migrating existing workloads can affect your operating model.

Understanding migration

Cloud migration is the process of moving an existing application, data, or infrastructure from one cloud service to another, or from an on-premises setup to a cloud-based environment. It can help organizations to improve their operational and cost efficiency and IT scalability while providing access to advanced services such as AI and machine learning or the **internet of things** (**IoT**). Common scenarios where cloud migration is used include the following:

- **Server consolidation**: This typically involves consolidating multiple physical servers into a single cloud platform, reducing overhead costs and improving scalability. The process involves migrating the data onto the new platform and ensuring that connections between different components remain intact. Additionally, it requires making any necessary changes to ensure applications run optimally in the new environment.

- **Disaster recovery**: By having backups of important data stored on a separate cloud platform, organizations are better equipped to recover quickly from outages or attacks since they can quickly move operations over to the backup system. Cloud providers often offer specialized disaster recovery solutions designed for this purpose that can help organizations minimize downtime and mitigate losses associated with any potential disruption.

- **Data migration**: In many cases, companies need to transfer large amounts of sensitive data between different cloud environments to make use of specific services or take advantage of new features offered by a provider, such as scalability or advanced analytics capabilities. The process involves designing secure connections between systems and then transferring all relevant information over a VPN using secure protocols such as SFTP or HTTPS.

- **Application modernization**: Hybrid cloud environments allow organizations to transition legacy applications onto modern cloud platforms with greater ease than ever before. By using container technology such as Docker, it is possible to repackage applications so they can be deployed across different clouds without any major changes being required the first time around, which helps keep migration costs low while still taking advantage of the improved performance these solutions provide.

- **Multi-cloud strategy adoption**: More companies are now adopting multi-cloud strategies instead of relying solely on one vendor for all their needs in order to leverage more specialized services from multiple providers at once while also benefiting from competitive pricing options available on certain services. This typically entails setting up connection points between different cloud platforms and then designing methods for securely transferring information back and forth between them when needed so that all applications remain functional despite running across different infrastructures at once.

In addition to these scenarios, there are various other benefits associated with migrating workloads and applications to the cloud, including improved flexibility, scalability, cost savings, accessibility, and security levels; however, each organization must determine what type of migration strategy works best for them depending on their unique requirements before taking the plunge into this powerful technology landscape. The Azure CAF can help organizations to assess their readiness for cloud adoption and build a tailored plan that helps them to successfully move to the cloud while ensuring they have the right processes, tools, and strategies in place. With this framework in hand, organizations can then rest assured that they are well equipped to achieve the best results from their cloud migration endeavors.

A cloud migration team structure is key to successful cloud adoption. It is important to ensure that the right personnel are in place to manage and oversee the migration process. The team should include three main roles – a project manager, a technical lead, and an operations lead:

- **Project manager**: The project manager will be responsible for setting up the organizational framework and timeline for the cloud migration. This includes defining objectives, managing resources, establishing milestones, monitoring schedules, and communicating with stakeholders. They must also have a clear understanding of the cloud platform being used so they can effectively coordinate between teams.

- **Technical lead**: This role requires someone who is highly proficient with the cloud platform being used and knowledgeable about the applications and systems. They will be responsible for developing plans for data migration, ensuring proper system configurations are in place, overseeing installation/launch activities, and testing services before deployment. In addition to this, they must stay up to date on changes or updates in order to keep everything running smoothly during the transition period.

- **Operations lead**: The operations lead will be responsible for overseeing day-to-day activities once all systems have been migrated over. This includes monitoring performance metrics such as latency and throughput as well as managing user access control processes to ensure that only authorized personnel have access to sensitive data or applications. They should also be able to quickly respond to any incidents or outages while also staying abreast of new features or services offered by providers so they can take advantage of these offerings when necessary.

In addition to these core roles, there may be a need for additional personnel depending on the complexity of the migration process, such as security experts or database administrators if these areas are affected by changes made during the transition process. It is essential that everyone involved understands their responsibilities and has access to all relevant information needed for success, so communication must be established at every level of the organization, from upper management down through individual contributors involved in each step of the process, from planning through to the completion stage.

Cloud migration can be a complex process, and there are many potential pitfalls organizations need to be aware of to ensure a successful transition. The most common cloud migration pitfalls include the following:

- **Unfamiliarity with the cloud environment**: It is essential that organizations have a basic understanding of the cloud platform before moving their workloads. This includes being familiar with its various components, such as storage, security protocols, and the computing power required to ensure that their systems can run well on it. Without proper knowledge and preparation, organizations may find themselves facing unexpected issues during the transition period.

- **Data privacy and compliance**: Organizations need to make sure that all data transferred or stored in cloud environments adheres to any relevant laws or regulations. Depending on the nature of their data, they may also need to take additional precautions such as encryption or access control measures to protect it from being accessed by unauthorized parties.

- **Insufficient bandwidth**: A slow internet connection can impede the progress of migration significantly as large amounts of data must be transferred between the source server and destination cloud environment. It is important to assess available bandwidth capacity beforehand to determine how many applications can be migrated reasonably and estimate the cost and time required for migration. Related to the first point, cloud providers typically have solutions to address this issue, but organizations need to be aware of these.

- **Security vulnerabilities**: Hackers may target vulnerable applications during the transition period, so extra caution must be taken when configuring systems for use in the cloud environment. Organizations should also consider implementing additional security measures such as multi-factor authentication or automated security policy solutions in order to keep sensitive information safe from malicious actors.

- **Poor planning**: Moving systems and applications over to a cloud platform requires careful planning if it is going to be done efficiently and successfully. Organizations should create a detailed timeline for each step of the process, including tasks such as assessing existing infrastructure, determining resource requirements, scheduling downtime windows, testing new services before deployment, and so on, so they can stay on track with their objectives throughout every stage of the migration activity without running into any unexpected delays or obstacles along the way.

The Microsoft Azure CAF provides organizations with a comprehensive approach to help them maximize the potential of their cloud investments. It is divided into five somewhat overlapping and iterative phases: *Strategy; Plan; Ready; Migrate and Innovate; Secure; Manage and Govern.* The Microsoft Azure CAF allows organizations to tailor their strategies based on their individual needs. By following best practices at each stage, companies can leverage cloud services such as scalability and agility while reducing the costs and complexity associated with traditional IT models.

Organizations should also consider utilizing tools such as Cost Management or practices such as DevOps to increase operational efficiency by automating various tasks associated with cloud service management. This will help them save time and resources, which in turn can be used for other important activities that contribute toward the success of their cloud migration. By building the right team and anticipating the common pitfalls, organizations can aim for a smooth transition to the cloud and get the best results from their investments in new technology and a cloud-first operating model.

Summary

A new cloud operating model in a well-designed landing zone allows organizations to maximize the potential of their cloud adoption investment. But it requires careful planning and preparation for migration, ensuring data privacy and compliance, strengthening security measures, and leveraging tools and best practices such as Azure Cost Management and DevOps. In this chapter, we barely scratched the surface of landing zones, so encourage you to explore the extensive resources available on this topic with the Azure Cloud Adoption Framework. With proper knowledge and preparation, companies can reap the benefits of scalability and agility while reducing costs associated with traditional IT models. We introduced the Microsoft Enterprise Skills Initiative, an extremely valuable knowledge resource – if your organization can access it, be sure to make the most of it. Overall, a cloud operating model may be familiar to those working in a traditional IT environment, but with many more opportunities and reasons to automate work. Developing cloud infrastructure automation to streamline cloud service management activities will save time in the long run and free resources to work on other important tasks that contribute toward the success of the cloud journey. In the next chapter, we will go deeper into cloud architecture, design patterns, and best practices and how the Azure Well-Architected Framework is a critical component of successful cloud implementations.

8
Architecting Your Cloud

For you to set your organization on the journey toward a successful Azure adoption, you must also understand the Azure Well-Architected Framework. We can call this framework the little sibling of the Azure Cloud Adoption Framework. In this chapter, we will go through the Well-Architected Framework briefly, but with enough details for you to be able to understand everything you need to know about it at a high-enough level to be able to guide your architectural design and the path your organization needs to take to successfully complete projects based on it.

In this chapter, we will cover the following main topics:

- Azure Well-Architected Framework
- Reliability
- Cost management and optimization
- Operational excellence
- Performance efficiency
- Security
- Architecture review

Azure Well-Architected Framework

You want to plan your design decisions, the technology you will use, and the way to implement those decisions and technologies into a coherent solution based on the following pillars of the Well-Architected Framework:

- Reliability
- Cost optimization
- Operational excellence
- Performance efficiency
- Security

Each of these pillars is always important – for every workload. However, your focus can be prioritized based on the specific needs of the workflow.

For example, focus on reliability in the following cases:

- You are building a public service that will scale quickly and violently
- You are responding to telemetry showing that reliability is an issue
- The team is having a hard time focusing on new features because they are hard pressed supporting the running services

Focus on cost optimization in the following cases:

- You have completed the lift-and-shift migration of workloads into Azure
- You are working on a project that wants to improve the marginal user cost due to business pressures
- You've been focusing elsewhere for a while and haven't done a regular cost optimization exercise for more than six months (Azure is constantly evolving – new services, new tiers, and new configuration options are coming out all the time and after six months, there will likely be new ways to optimize your existing resources).

For example, focus on operational excellence in the following cases:

- Services are struggling to stay up (reliability) due to a lack of automation
- Services have been multiplying in the past six months, the teams are struggling to support them, and you are approaching linearity in the people-to-services mapping
- You've got new folks rotated into the team and having them focus on operational excellence topics will give them a great insight into how the landing zone and services are running

Focus on performance efficiency in the following cases:

- You are expecting load increases by orders of magnitude more than usual (for example, in December for the holidays) so you can test, double-test, and check and discover all the performance bottlenecks that may not have been tested before to such a degree
- A team is struggling to interconnect gracefully with other services – internal and external
- There are any (explained or unexplained) performance spikes in your telemetry and you must work to smooth them over

Focus on security in the following cases:

- You must *always* focus on security (assume breach, bake security into everything you and the teams do, and focus on zero trust)

Let's just parse that last thing:

- If you haven't been focusing on reliability, your services may be down some of the time. And that is bad, yes. But at least some of the time your services are up and running.

- If you haven't been focusing on cost, your services may be more expensive than they should be. But hey, they are up and running.

- If you haven't been focusing on operational excellence, your teams are struggling and hate you, but what's a little hate among friends?

- If you haven't been focusing on performance efficiency, your customers hate you as everything they do with you is *sloooow*, but at least the services are all available.

- And, if you haven't been focusing on security, you have been hacked. You might or might not know it, but you have been hacked.

Well, now that we dug into these terms, it kind of looks like you need to focus on *all the pillars*, or your business will be in trouble. Hopefully, you've learned the lesson here and not "on the job." You are welcome.

Let's recap:

- Focus on reliability for your services to be available and serving traffic so your customers don't abandon you for a more reliable service provider

- Focus on cost optimization so you don't bankrupt your organization

- Focus on operational excellence so your teams are not always firefighting and neglecting feature work

- Focus on performance efficiency so your applications don't resort to showing progress bars for every operation (frontend developers tend to add UI elements to improve the user's experience of a slow application) so your customers don't abandon you for a faster provider

- Focus on security so you don't have to pay ransoms, fines, and, worst of all, brand perception and reputation penalties, so your customers don't abandon you (in the best case) and sue you immediately after (the second-best case)

OK, so you are now ready to focus on all the pillars of the Well-Architected Dramework! Excellent.

Let us now focus on how to focus…

In order to know how to achieve the best outcome out of each of these five pillars, you have to know which questions to ask, what decisions to make, and how to organize your and teams' time to deliver the best result with the least amount of micromanagement.

Your services need to meet all the commitments your organization has made to its customers – internal or external, developers or end users. Reliability is achieved only by architecting it into all your services.

Some things really need to be done now, not later

You *can* still add on reliability after the fact, but usually at the cost of effort, and it is always a pain. I've yet to see a non-reliable architecture made reliable easily.

This is advice investors always give to start-ups – don't focus on reliability (sometimes they call it scaling) early on. Create a quick-and-dirty proof of concept and minimal viable product, which then always ends up being presented as v2 and goes straight into production. While I cannot argue with the business sense of that – after all, very few start-ups achieve market fit quickly, or at all, and need scale, so no sense baking it in.

Make sense?

No!

That used to be good advice because reliability used to be hard. With the right architecture and the right cloud services at the correct tiers, reliability is laughably easy (I, of course, welcome your feedback here – "Um, actually… reliability isn't laughably easy… because once I…"). But it must be an architecture decision from the start.

To contrast that with security. Both reliability and security must be baked in. However, security is orders of magnitude more difficult to implement and do well and stay on top of going forward. Reliability is just a set of patterns you can always follow and you are good to go – if you follow them. A lot of improvements once the workload is in production will come from SREs anyway. But they will have an easy job if reliability was not an afterthought.

Initial investment to understand the reliability patterns is miniscule and plan to bake them in everywhere. There is quite a bit more effort required to communicate that to all the teams, but a lot less than informing them after the fact.

So, how do we start doing this?

Make one decision now, for all services. All services must run in an active – active configuration from day 0 across two Azure (paired) regions. This will force reliability into all services from day 0, and while your job won't be complete there, it will be so much easier. And if this is your only decision on reliability, you are well on your way. This one decision will force comprehensive health checks, automated deployments/rollbacks and failovers, high-availability considerations, data replication, and great observability from day 0 – by teams rushing to implement Azure paired region deployments.

What questions to ask

These are some of the important questions you need to ask, and some guidelines as answers.

What are the reliability targets and metrics you will define and implement for your services?

Two things here. Firstly, don't make your services more reliable than they need to be. Remember five 9s availability? Each additional nine of availability is prohibitively more expensive to achieve. Also, services that are too reliable are a problem in and of themselves (check back to the *Site reliability engineering* section in *Chapter 6*). Secondly, make your services as reliable as the business needs them to be. There is often a plan put into action that prepares tiers for reliability and then all services must fit into one of the tiers. And they are usually tiered by a person who's never even heard of statistics, so all services end up over- or under-provisioned. Work with the business to the exact standard each service needs to be reliable to. And remember, not all services are real time; some are near real time, and some are nowhere near real time – so the needs for reliability will be different. For example, sending an email notification in almost all circumstances is not a real-time function. That means that the send email service might not need to be the most reliable service; you can always retry in a few seconds, minutes, or hours, and sometimes even a day isn't an issue. So, really push back on business decisions around reliability that make sense to each service. Thirdly, know your **recovery time objectives** (**RTOs**) and **recovery point objectives** (**RPOs**).

Are your services resilient? What happens when the service fails? What happens when a dependent service fails? What happens...

Resiliency is everything from deploying services across Azure regions and ensuring all single points of failure are removed to identifying fault points and modes (as in, different ways the service can fail) and planning for the graceful degradation of a service in case it starts getting into trouble (for example, if the service has a dependency on a third party service, what can you still deliver if the third-party service fails?).

What are your scaling and capacity requirements? Which Azure services do you need in each region you are running in?

Although the cloud capacity can be perceived as essentially infinitely scalable, there are some requirements that might make you aware of just how much the cloud isn't infinite. For example, if you need 200,000 VMs with a particular GPU size for your rendering needs in each of the 8 regions, you might need to plan for that with Microsoft.

How is your service architected? How is the data stored?

This question is all about ensuring your application processes are stateless, sessions are not sticky, and data storage is externalized and redundant, as well as understanding how load balancing and health checking are implemented and so on. Are your calls direct or are they decoupled, is autoscaling automated, and is client routing efficient?

Hopefully, the correct patterns are clear from the preceding paragraph.

Have you considered security and scalability and which reliability allowances you need to make?

For example, are all endpoints secured? Remember, they are not distributed or redundant and there will usually be more than one of them. Sometimes, there might be hundreds or thousands of endpoints because of reliability decisions.

Another example is emergency break-glass accounts. Are they redundant, have you tested the process for their activation, and are they secure enough that they will be an actual break-glass account and not just another compromised account?

How will you monitor and automatically assess the health of a service?

Ping is great when debugging, but it is not sufficient for the comprehensive monitoring of the health of a distributed service. A service has dependencies on data stores, other services, and other third-party services, so your service being available to answer a ping is not sufficient. So, what should your service health checks do? And don't forget to monitor your long-running processes for actual completion and any issues there.

Asking questions is fun, and uncovering even more questions can make you feel like you are adding value, but very soon, you will need to find solutions and make some decisions.

What decisions to make

As demonstrated with the preceding sample questions, you will need to make a lot of these decisions in the architecture stage and then communicate them to your teams.

Consider your list of non-functional requirements and consider what your red lines are! For example, you might let slide and prioritize later to fix a sticky session for a less-used and less data-complex service, but you might not compromise on the multi-region requirement. Consider these red lines and priorities carefully.

How to organize your teams

Your teams need to understand well the business requirements around reliability and understand and accept non-functional requirements and red lines in their planning sessions or as part of their tasks. Hopefully, documenting changes is a normal practice of your teams. The same goes with requirements coming from these architecture decisions and they should be internalized into the teams' everyday practices.

Cost management and optimization

When talking about cost optimization, consider it a modeling exercise where Azure resources get mapped to your organization (or across the levels of your organizational hierarchy). In order to estimate the cost in advance of deploying a workload, confirm the cost when running in development and production and monitor and stay on top of those costs. Finally (if that is how your organization works), apportion the costs to different parts of your organization.

Which questions to ask

These are some of the important questions you need to ask, and some guidelines as answers.

- How are you modeling the costs?

    If you want a recommendation here, it is to always model your infrastructure costs in the Azure calculator (as opposed to your own internal Excel sheet) because it will be easy to collaborate on them, the costs will be updated live, and you can easily export, change, or import them when needed. Staying in sync with your own custom Excel sheet (or a another tool) is not wise.

    Is the price model clear for everyone? Is it pay as you go, easy to understand, transparent, and clear on which costs are initial and which are reoccurring, as well as which costs will increase with the number of customers or increased load?

    Are any dependencies understood as well? For example, are there dependencies on shared services from the landing zone?

    Have you considered all the different ways that Microsoft discounts Azure services and software licenses – volume, pre-purchase/reservation, partner programs, and so on?

- How will you monitor ongoing costs?

    Have you integrated Power BI with cost management in Azure? Can the finance team access it and get direct insights? Or do you have to query and prepare results for them? Can everyone in the organization see the costs, especially costs apportioned to them or their part of the organization?

    You will need to set up budgets and alerts as well and define owners for each alert. Are all costs placed in a single location or do you have to hunt for them across workloads, environments, organizations, and so on?

- Have you thought about your delivery teams and the SREs?

    Is the deployment of new versions automated? If you are doing blue/green deployments, are you deleting old deployments once they are no longer needed? Are there differences between development and production service tiers? Are you shutting down services when no longer needed?

    Do you know the ratio of cost of production versus non-production environments? Do you know the marginal cost per customer, and is that increasing linearly or decreasing exponentially (hint: the latter is better)?

These questions will help you gauge the maturity level of your cloud governance model, specifically around how well your organization is positioned to manage costs.

What decisions to make

You need to know how much you are spending on the cloud. You also need to know how much you are spending on each service, each Azure service, each environment, and so on. This includes things such as data storage costs, operational costs (including the teams themselves), third-party tools and libraries, consultancy services, and so on.

You must make decisions that make operational sense, but also business and common sense. For example, how do you apportion costs and inform stakeholders how much they are costing the organization?

Another important decision is when you will step in and at what point the teams will be made aware of issues around costs!

How to organize your teams

The best way to organize teams around the cost optimization topic is to make them responsible for costs, especially non-budgeted costs. No need to annoy them if the costs are within budget and cost optimization is part of their day-to-day work, but if it isn't, that is when responsibility kicks in. Teams (and subsequently you) must always know the cost of running their services, including all the auxiliary costs mentioned earlier.

Teams must be aware of costs and the need to rearchitect to optimize costs. The goal is not to squeeze the team on costs (for example, making them compromise on realistic reliability), but it is to stop unnecessary waste. Food waste is incredibly bad (some would say immoral), but people shouldn't stop eating. They should try and minimize waste as much as possible. The same goes for teams running a service in the cloud. Understanding the fundamentals of cost attribution, control, and optimization is essential. This means ensuring operations are not disrupted and costs are managed and reported automatically, rather than from ad hoc queries.

Manage costs by budgeting, analyzing forecasts, reserving instances, and tracking actual spend. There is a distinction between Azure budgets and your company budgets. Azure budgets should be set at a reasonable level to stop excessive alarming. Your company needs to move beyond rigorous financial planning for the cloud into the agile world of cloud operations and cost optimization.

Did you know you can create alerts/tickets for teams based on the automated evaluation of cost on the environments they are running? Use that! If costs change in any significant way, create a ticket automatically and let the team own and align on a resolution within the organizational parameters.

No one should be watching graphs, counting euros, and answering ad hoc cost questions. Planning is the key to scaling cost management. Only the teams themselves should check out graphs, count euros, and answer questions on cost – when it makes sense for them to be able to debug an issue.

Operational excellence

Automation is key – both to the hearts of the authors of this book and to operational excellence. The less manual work required, the better.

There will always be things that haven't been automated yet! And here is some guidance on what operational excellence means and how to achieve it.

> **Operational excellence = happy teams?**
>
> The second-most-often-mentioned reason for team dissatisfaction (in my opinion) is failures in operational excellence. The number one reason is, of course, money. And reason zero is, of course, always the manager.
>
> If the teams are failing at operational excellence, it means tasks are boring, unproductive, repetitive, or – worst of all – something has been raised again and again as an issue and the organization has ignored the complaints. Or maybe there are too many emergencies and so the focus is on firefighting rather than automation and the development of new features.
>
> So please, do better and you will have a lot more satisfied team.

Which questions to ask

These are some of the important questions you need to ask, and some guidelines as answers:

- As always, roles and responsibilities! Have you planned those out?

  Are developers, operations and cloud platform teams, SREs, and who knows who else in your organization clear on the processes, and who is responsible for each part? Of course, development and operations should be the same team, à la DevOps – ideally implemented following the SRE team pattern.

  Once the roles and responsibilities are planned out, have you planned for the remediation of issues, and have you tested them often and in the actual (replica, preproduction, etc.) environments?

  Are bug fixes prioritized ahead of the development of new functionality? Is prioritization generally clear and well understood?

- Are you documenting everything?

  This includes all the dependencies documentation, in other clouds or on-premises as well. Third parties as well? DNS, CDN? Email?

- Resiliency and self-healing! Are these words important to you?

  They are just synonyms of another word you should be familiar with – automation! Services will fail in a variety of ways (and ideally, you know this and have comprehensively tested the service under multiple scenarios). To achieve resiliency and/or self-healing, you have options:

  - Scale – if you are having performance issues – then deploy more instances to handle load

  - Redeploy the last-known good version – there is something wrong and you may need to take a step back (literally) and assess the situation

  - Minimize the available functionality to only essential functions – stop returning detailed statistics to users to minimize traffic across the network while sending detailed telemetry to operations

  - Fake responses – no need to update the high-scoring table every second; send back a stale version

- How are you managing service configuration?

  Are you on top of new functionalities coming from Azure and taking advantage of them? Are you storing the configuration in a managed location? Is soft-delete (history) available for configuration? Are you rotating secrets?

  It is important that the configuration is known, committed next to code and infrastructure as code, available, and documented, and can be easily accessed and changed by those who need to do so. Perhaps a highly available central service for configuration access would be a good idea for your services. Azure services usually provide a usable configuration location for themselves.

What decisions to make

Automate everything. If you take away one thing from this book, let it be that. If it can be automated, it must be automated. You can only scale through automation. So, how do you best scale different services in Azure? The bad news is there are so many services and ways to scale in Azure! The good news is none of it is hard if you know what you are doing!

How to organize your teams

This one is easy. Focus on automation. Making improvements in CI/CD is the teams' daily task. Make sure you stay on top of this, as there are two types of people – those who automate and those who do the same repetitive tasks over and over again without automating them.

> **Some people like repetition**
>
> Personally, I recommend keeping the folks that automate everything and suggesting a move to security pen-testing for those folks that enjoy repetitive tasks. Attempting a breach over and over again (in ever so slightly different ways) is what these folks are good at – aside from manual quality assurance testing, which you should be doing less and less over time as you automate that as well.

An easy way to track this is by tracking manual tasks in small categories (as one trivial example, a password reset) and then seeing how many times a task of the same category gets done. As with everything, however, there is good and bad automation.

> **Certificate management**
>
> Automating the rotation of your certificates is bad if the automated process runs once every two years. A good approach would be to automate it every quarter. The best approach? Daily! You want the automation to run often to verify that it is running well and so that when it fails quickly, you can fix it. If certificates are rotated daily, you will know rotation works and in two years' time it will still work – because every day between now and two years from now, it has worked fine. This is one of those counter-intuitive patterns. As usual, the motto is – if it works, don't touch it. The problem then is that one time it fails. So, make it work every day.

Performance efficiency

It's all well and good having a reliable and secure service, but the performance must also be acceptable to your customers. The only thing worse than the service you are trying to use being down is it being available but lacking in performance.

We live in a world of instant gratification and a huge number of options available to us – and so any delay standing between us and what we want is unacceptable. We've built an amazing world based on revolutionary technology and therefore, it is unacceptable that your service isn't performant.

Instant response is a utopia – or is it?

In 1968, Robert Miller published a paper titled *Responsible time in man-computer conversational transactions* (https://dl.acm.org/doi/abs/10.1145/1476589.1476628) in which he claims that a response time under 100 ms is perceived by people as instantaneous. Great! We can do a lot in 100 ms (ms = milliseconds; there are 1,000 milliseconds in 1 second).

So, as a rule of thumb, if you can get all your responses to be under 100 ms (including the user-service internet latency), you will be able to deliver one of the best service response times in the world (or perceived as such). While 100 ms is a lot, it isn't nearly enough if your service crunches data, calls third-party services, formats the response, and so on.

So, the number-one rule in achieving sufficient performance through architectural decisions is to minimize the number of calls beyond the initial call. If your service needs to coordinate dozens of other services, and compute and store data, you will need to find a better way. Remember, not all services need to be in real time or near real time.

In gaming, 50 ms is the target performance to reach. Humans cannot react or detect anything below 13 ms though. Great! The service has 13 ms given to it by our human; 50 ms if it is used for competitive gaming, and about 100 ms to be perceived as instantaneous in all other circumstances. These are your targets, and it is your responsibility to reach them.

> **Life on the edge**
>
> I am lucky as I live in Dublin (Ireland). Often, gaming servers are co-located in the cloud and Dublin is so close to data centers here that I often have near 0 ms latency. I am not bragging; I am saying that it is possible to achieve this, but you may need to bring your computations closer to the customers, to the so-called edge.

Which questions to ask

These are some of the important questions you need to ask, and some guidelines as answers.

- Performance matters! How can your architecture plan for it when you have customers all over the world?

  The first thing to know is that everyone's internet speed is different, and speed can vary between workloads, times, the weather, character encoding, encryption, and so on. That is also good news for your services, as the background latency of the internet is working for you. If your services are underperforming a little bit, the issues will be hidden by the background internet latency.

  Aside from that, how do you improve your **time to first byte** (TTFB), an important performance metric in web application development? You can always deploy across multiple regions! You can deploy to paired regions and beyond. With Azure, you can deploy to more than 60 regions! Is any of these regions sufficiently close to your primary user base? If not, you can use Azure CDN services, which are edge locations ready to host your content, so it is even closer to the customers. Then, it is a question of using third-party services to further expand your CDN locations. In the enterprise world, there are dedicated lines that help avoid internet noise.

- How predictable is the usage of your services?

  Autoscaling is great. The recommended strategy with pay-by-the-minute services is to autoscale out generously, perhaps 10x (so, going from 1 instance to 10 or from 10 to 100 and so on). This ensures minimal disruptions and helps with even the largest traffic volumes. If you are aware of marketing activity in advance, scale in advance as well (automatically and/or manually, as needed). Then, pair this with autoscaling so you can reign in the costs.

The more predictable your service load is, the easier this gets. Horizontal scaling is key – aim for multiple instances, not larger instances. Test autoscaling under heavy load and measure the time it takes to scale an instance in and out. In Azure, consider the limits and quotas, especially on computer instances that are deliberately very small by default (to protect you from exploding costs). This goes hand in hand with planning early. You might need hundreds of thousands of instances. Talk to Microsoft now, not when your autoscaling fails. You can see your quotas live in your subscription in the Azure portal.

There are other traffic management methods, such as priority queues (for your customers) and non-priority queues for everyone else (for example, Reddit serves a read-only (mostly) static site to everyone not signed in). If you are not selling your services in some markets, consider blocking traffic from those markets altogether. Consider these if you can! Once things go bad, who are you going to call?

The answer better not be Ghostbusters! You need to have a plan in place in case you need to get Microsoft support for an issue you are experiencing. Choose your Azure support options well in advance to having a need for them!

- Have you had issues before?

Firstly, why haven't you addressed them? There really is no excuse for facing the same issue twice. Consider how issues you've had before were resolved, document them, and make sure everyone is familiar with them. You should have a solution for common issues (automated, hopefully) – such as storage not sufficient, CPU usage is high, and memory is at 100% utilization. You also need to be familiar with monitoring, alarming, and the profiling tools. You cannot be searching for the right tool; you must already have it and know how to use it.

If you need to make a change, make a drastic one. 100 VMs locked? Kill all 98 or 100! Don't kill 1. You need to resolve an issue quickly and you can only observe the change if it is a drastic one. You will know immediately whether it has helped or not. Have in place a way to split traffic if you need to and have failover URL alternatives for each service so you can slowly recover with slowly increasing traffic. If you have a billion requests a second failing, there is nothing you can do (well, there is, but if you are reading this book, you are likely not one of those people who know how to handle it) to redirect all of it to a secondary location and not have it kill the secondary location as well.

You should have all endpoints under a throttling management solution (in Azure, one option is API Management)! Also, ensure your own services coming out of graceful degradation modes do not immediately hammer your other services. Fail quickly, recover slowly is the rule for services.

Your CI/CD pipelines must be familiar to your teams and work flawlessly because unless the issue is with the CI/CD pipelines, you should never make manual interventions in production (unless you are forced to make drastic decisions). Your fix – it goes through CI/CD, and your tests should scale (in emergency mode) so tests that under normal circumstances take 2 hours to complete now complete in 60 seconds or less.

I know a lot of this sounds like reliability issues, and they are. Usually, though, system performance will be your first indicator that something is not right.

What decisions to make

Benchmark your workloads. Know the breaking points of services. If the service works with 1 billion connections, try 10 or 100 connections. If the service works when all dependencies are 100% fine, know what happens when dependencies are underperforming, missing, or malicious. Then, plan for it all.

When planning, plan for general solutions – solutions that can mitigate many issues. Do not plan for Hollywood scenarios and prepare specific solutions that only work in this one unlikely case. Also know who your allies are – law enforcement agencies, industry experts, consultants, and even competitors. Know who to turn to.

How to organize your teams

Your observability around performance must be flawless if you want to catch reliability issues before they escalate. Once something goes wrong, expect other things to go wrong as well in the days, weeks, and months after the incident. Plan your staffing accordingly. Trouble never comes alone.

> **Expect the unexpected**
>
> COVID-19 caused disruption that we couldn't predict with sufficient accuracy at the time. It brought to the forefront logistics issues we didn't know we had and that we will have to deal with for years to come. It made accidents such as blocking a ship route more catastrophic than they would have been had they been the only thing to deal with. It brought opportunists closer to executing plans where they might have hesitated before.
>
> We can use the same analogy to see how many things happen when one issue occurs. Other issues surface; issues that would previously have been minor now become major and opportunists will be ready to take advantage, which should encourage you to reconsider executing anything through a different process than before. If a patch has been provided through CI/CD in Azure DevOps, do not accept a patch over email or from a third-party service you've not used before in the normal course of your work (unless, of course, that usual path has an issue. Even then, verify again and again whether people are who they say they are and are sending you the correct files through correct channels).

Hopefully, you understand now that performance is the canary, your best canary. A team of cats equipped to deal with canaries is what you need, so the exact opposite of what you saw in cartoons.

Security

As more and more businesses move their operations to the cloud, ensuring the security of data and applications has become a critical concern. Cloud security refers to the set of policies, technologies, and controls that are put in place to protect cloud-based systems from unauthorized access, theft, or damage. With sensitive information being stored in the cloud, it's essential for companies to have robust security measures in place to safeguard their assets and maintain trust with their customers.

In this era of digital transformation, cloud security is an increasingly important aspect of an overall cybersecurity strategy.

You, as an architect, need to start by asking questions to gain a better understanding of the risks involved and how to mitigate them, identify gaps in current measures, and provide insights into areas that need improvement. This can help you make informed decisions about your organization's cybersecurity strategy.

Which questions to ask

There are so many questions to ask around security that we would need a separate book, or two, or nine, or even fifty to cover them all. You can start with these.

Do you know what threat analysis is and can your teams do it?

Threat modeling identifies threat vectors across the board. Everyone in your business, not just the technical folks, will need to participate in identifying threats, ranking them based on the business impact, mapping them to mitigations, and communicating all this to your entire organization. In security, everyone is a stakeholder.

Start of your triage process must be quick and efficient and with clear responsibilities beyond triage.

Security requirements are defined and obeyed by all teams, and then your red/blue/purple teams verify them in action. Whenever possible, use existing frameworks to define requirements so you don't accidentally miss an area of focus. Classify business critical services and prioritize that analysis.

It goes without saying that all this needs to be done before the services go live.

Are you in a regulated industry?

The good news is there should be no difference in approaches across regulated and unregulated industries – in theory. Everyone should follow all the security best practices, right? Well, the bad news is no one does, at least not completely, hence the regulation. So, regulation of security is good, as you can usually follow industry best practices and add on any other security requirements beyond the defaults in your industry.

Just remember that standards of compliance aren't there to help you secure your workloads; they are there to remind you that you are required to do things a certain way and prove that you are doing them. You can do all of them and still not be secure. So, take heed.

Regulatory and governance requirements must be understood and followed and reviewed and performed regularly. For all Azure services, you can also find security recommendations in the documentation.

Always cover the basics (such as **Open Web Application Security Project** (**OWASP**)) as that raises the bar and the amount of effort to find and exploit vulnerabilities from drive-by attacks to much more targeted and sophisticated attacks. A sufficiently knowledgeable, persistent, and funded opponent will always get in no matter what you do. Your ideal goal is to raise the effort so high that it just isn't worth it. Your second-best goal is having observability so good that it gives you time to detect and act before they get in – and gives you time to call for assistance from law enforcement.

But since it is almost guaranteed a security incident will happen, practice a response to everything from "we've detected them early" to "they are selling all our data on the black market and making us look bad on social media and in the news."

One final word here – insurance.

Are you baking in security or bolting it on?

Everyone must think about security – from before the first line of code to the last request before retiring the service. This is the only way, and yes, continuous education is key. Ensure security teams and peer reviews are involved every step of the way.

If you can manage it – have a bug bounty policy.

Rotate all your secrets, often – at most, daily.

Public services, yes?

Absolutely. Whenever necessary! For example, serving your UI might be done from a public service, but accessing services, storage, and the platform should only be published as private endpoints.

Luckily, in Azure, (almost) all services (for example, Azure Storage or Azure CosmosDB) have an option to only have a private endpoint! The ones that don't are usually those that are public by their nature, such as Azure Traffic Manager.

Overall, asking questions about cloud security is essential for maintaining a secure environment in the era of digital transformation. It helps organizations stay ahead of threats, comply with regulations, build trust with customers, and improve their overall cybersecurity posture. Customers trust organizations that take their security seriously. By demonstrating a commitment to cloud security through asking questions and implementing strong measures, you can build trust with your customers and stakeholders.

What decisions to make

You likely cannot make the right decision around security. Decisions on security, being one of the absolute priorities, must come from the very top of your organization.

All other decisions will be yours and your teams. What kind of organization do you want to be?

> **Compromise**
>
> You will be forced to make day-to-day decisions around security, and every day there will be a request to compromise on security – by people wanting to move quickly with their small tasks, people who haven't properly considered security, or people who just don't care.
>
> It takes an amazing level of strength of character and conviction to say, "We must stop, pause, and think," every time – to developers, commercial folks, an SRE during an emergency, the CEO, and customers. But you must. I promise you; you will rarely be forced to say no. There is always a way to have security baked in and achieve business goals. Compromise once and you are compromised forever.

How to organize your teams

Everything from brand reputation to data security is impacted by your security (counter) measures. Will you know if there is a data breach? What will you answer if the CEO asks whether you are secure? Or if they ask whether the company is going to be a headline in tomorrow's news? How do you assure them security is considered at every step of the way and ensure that that is in fact the case? What tools are there for you and how can you use them to their maximum potential? This is what you need to plan for.

Just know that security is best done before an incident, not during one. Do have knowledgeable red/blue/purple teams and have them test the security of everything and everywhere – repeatedly.

> **Try something different**
>
> Bring in third parties regularly, swap team members between red/blue teams, and try weird and crazy things that would "*never work.*" You'd be surprised how often they do, in fact, work.

Architecture reviews

A great start when architecting a workload is the Azure reference architectures page: `https://docs.microsoft.com/en-us/azure/architecture/browse/`.

Your workloads are special, but it is very unlikely you are doing something never seen before, so check and see whether one of these can give you a jump-start toward your end architecture.

So, how do you perform architecture reviews?

The intent behind architecture reviews of reliability, cost management, operational excellence, performance efficiency, and security is to ensure architecture is unified, non-functional requirements are considered and implemented, and everyone is on the same page.

You will need to plan the cadence, which, depending on the speed and the volume of change in your organization, might be from every two weeks to every quarter. Err on the side of two weeks; you can always cancel the review if there are no changes.

Here is an example of a quarterly plan from one SRE team (as an example – your needs will vary):

- Week 03: Capacity planning (for this quarter and the next) – with architecture, engineering, customer success, and marketing

- Week 04: Security review

- Week 05: Architecture planning (long term) – with architecture

- Week 06: Reliability review

- Week 08: Operational excellence review

- Week 10: Performance efficiency review

- Week 12: Cost optimization review

Whether you like this plan or not, note that this is just a way to get everyone together to force a discussion. Regardless SREs should be evaluating the need for formal discussion regularly or as event occur and can call for a meeting as needed. Agility, yay!

Set up an architecture board where everyone is welcome to come with their issues, suggestions, and review requests. Everyone should try and understand the existing architecture. Then, you can brainstorm improvements or new services required to improve the architecture pillars. Discuss any new business requirements and see whether they can be satisfied by the current architecture. If not, improve. Decide how to proceed and which team is responsible for the initial change and how they will work with other teams to bring them on the journey of improvement once they have completed their changes.

Document the reviews and have them freely available to everyone in the organization, and if you are so inclined, share them with interested customers (remove any information that might be exploitable – such as internal endpoint URLs).

Evolution

These are all continuous processes! They are here to stay.

There will (very likely) need to be a cultural change in your organization to accommodate all of this. If your teams are truly agile, it shouldn't be a big change. If your teams are pretending to be agile, there is a lot of work to do.

Business processes will have to evolve as well to accommodate improvements. Take this example:

- **Reliability**: May necessitate contractual changes to allow for the geographical replication of data

- **Cost management**: May require changes in organizational budgeting, tracking, and spending

- **Operational excellence**: May require team structure changes to accommodate the agility and independence required to make decisions quickly and be able to execute them

- **Performance efficiency**: May require a mindset change to understand the reasons behind requiring improved observability across the boards and may require better collaboration between teams on how to achieve it wholesale

- **Security**: May require organizational changes and reporting line changes to ensure security decisions cannot be rushed, overridden, or ignored

Is your organization ready for these changes? Can it execute them quickly?

It also requires a huge degree of self-reflection and acceptance of the need for further education across the organization. Balancing innovation on these architectural pillars and work on public-facing features will need to be considered. Who has priority? When? What are the red lines?

Do all services have the same priority or are there some that are mission critical? If there are mission-critical services, how do the processes differ across each of them? The traditional monolithic approach to building applications has given way to a more modular, microservices-based architecture that allows for greater flexibility and scalability. This evolution has been driven in part by the rise of cloud computing, which provides easy access to resources such as storage, processing power, and networking. In addition, advances in containerization and orchestration technologies such as Docker and Kubernetes have made it easier to manage and deploy these services at scale. As cloud technology continues to evolve, it is imperative that organizations keep up with the latest developments and trends in order to stay competitive and maximize the benefits of the cloud.

Azure is changing

Azure is constantly changing, evolving, and improving! Do you have a culture of continuous learning? If yes, do your teams and people have enough time for education? Are they encouraged to learn, or are they pressured to deliver? Once they learn something, how long before they can implement the new learning in production?

Once you've assessed a workload against a particular well-architected pillar and given your recommendation, you need to quickly figure out how to integrate the recommendation into the workflow of the responsible team. The team then needs to triage the backlog, plan the work, and execute on the recommendation. All through this, you need to monitor progress and be ready to step in and advise.

This can only happen if agility is high among your teams. If they are forced to execute on an 18-month roadmap, will your recommendations go to the back of the line? They shouldn't. How can your organization cope with that? One way to do this is by working closely with Microsoft Azure, who can offer guidance on how best to leverage their services and optimize performance. Regularly assess your current cloud infrastructure and processes and look for areas where improvements can be made or new technologies implemented. By staying agile and adaptable in the face of changing cloud technology, organizations can position themselves for long-term success in an evolving digital world.

Summary

Architecting your cloud involves applying design principles and best practices used to build and manage cloud-based systems. It involves optimizing the infrastructure, applications, and data for performance, security, scalability, and cost-effectiveness. The Azure Well-Architected Framework is a set of guidelines developed by Microsoft to help organizations design and deploy their cloud solutions on Azure in a secure, scalable, and reliable manner. The framework consists of five pillars: cost optimization, operational excellence, performance efficiency, reliability, and security.

Each pillar includes a set of best practices that organizations can follow to ensure they are following industry standards and achieving their desired outcomes from their cloud solutions. By adhering to these guidelines, organizations can reduce risk, optimize costs, improve performance, and enhance their ability to scale as needed. Overall, a critical component of good cloud architecture is the Azure Well-Architected Framework, which helps you leverage the benefits of cloud computing while maintaining control over infrastructure and operations.

And with that, we are almost done with the book. It is time to recap what we have covered throughout the book in the following chapter.

9
Closing Thoughts

Thank you for reading this book, you absolute legend.

Adopting cloud technologies can be a transformative journey for any organization. However, it's important to keep in mind that this transition will likely take some time and effort. In order to fully take advantage of Azure and other cloud platforms, it's important to have a clear understanding of your goals and objectives. Additionally, you'll need to make sure that your existing infrastructure is compatible with the cloud. With careful planning and execution, your organization can reap the many benefits of the cloud. We (the authors) started our journey on writing this book from a technical perspective but soon realized that we can provide more value through philosophy and anecdotes, so we expanded those.

Technical information is generally available all over the place, namely in cloud providers' own words, on their own web properties, and in other valuable books from Packt. But the insight you get from years and decades (has it really been that long?) of consulting with clients, leading software engineering teams, and directly managing cloud adoption programs have crystallized a few ideas that are crucial to the topic of this book.

Let's recap what we covered.

Reviewing Chapter 1, Introducing ... the Cloud

We started by introducing the cloud, that mythical creature that is always in the sky, flying about, and destroying people and organizations that are not ready for it. Oh, sorry, we're thinking of dragons there. Let's try again.

We started by introducing the cloud, that wonderful elusive concept, and – even worse – the concept that seems easy to understand but when put into practice or *adopted*, still seems to cause a lot of trouble for people and organizations trying to conquer it. Furthermore, we defined the target audience for this book – namely, architects, and specifically, cloud architects. While we tried to make the information available to all, of course, cloud architects and engineers are the ones to benefit from the cloud adoption and will lead cloud adoption. If you are working for one of the lucky organizations where the business and product and sales teams are also onboard for cloud adoption, then the information might be for them as well.

We also described ourselves (the authors) so you know who is preaching to you from up on high. There were some assumptions defined, as always when writing a book.

We also showcased the convention of the asides we'll have throughout the book that hopefully allow us to bring our opinions to the forefront for you, our experiences for you to judge, and our struggles to help you not make the same mistakes.

We briefly touched upon the five pillars of the Well-Architected Framework.

Now that we reflect on this, there was a lot of ground covered in this chapter. Hopefully, this chapter also showcased the due diligence required when reading – by searching and researching and looking for more information as you go along. Learning about the cloud can seem daunting and overwhelming, especially for those with limited previous knowledge or experience. However, by taking an incremental approach and focusing on specific topics, it is possible to bridge any gaps in understanding. Start by breaking down the subject into smaller chunks, such as exploring different components of cloud computing or researching specific use cases. Once a basic level of knowledge has been established, identify any additional gaps in order to get a clearer picture of what is required to understand the book better. Looking at further resources such as articles or videos that provide more in-depth information can be a great way to gain new insights and fill these knowledge gaps. Additionally, practicing tasks related to cloud computing will allow you to apply your theoretical learning directly and gain experience, which is invaluable when reading the book.

Reviewing Chapter 2, Adopting A Strategy For Success

Adopting a strategy for success – this should be the title of every chapter of your life, dear reader.

However, in the narrow confines of this book's topics, we tried to start you correctly on the strategy for success in adopting the cloud in your organization.

We mentioned the **garbage in, garbage out** (**GIGO**) strategy, which seems to be widely adopted by organizations, consultants, politicians, and yes, the public as well. Of course, we mentioned GIGO in the context of measuring success and Goodhart's law.

We also cautioned you about focusing too much on tech and not enough on business results. For example, Kubernetes is great, but business services delivered quickly are better and more valuable – and if Kubernetes must be the delivery vehicle, so be it. But start with the business. When implementing a cloud adoption strategy, it is important to be aware of the different business needs and problems that can arise. There can be a variety of factors that impact the success of cloud adoption, including budget constraints, security and data risks, availability of resources and personnel, as well as ethical considerations. To ensure that all stakeholders are on board with the proposed solution, it is beneficial to conduct research into your available options and articulate the benefits of each in clear and tangible terms: identify what value will be gained from a particular option. This could include cost savings, increased efficiency, improved customer experience, or any other gains that are likely to arise from taking the chosen path. For example, a cloud adoption strategy might provide cost savings due to reduced need for physical hardware or result in faster response times due to improved scalability and availability of resources. Additionally, depending on the scale and complexity of the cloud application being developed, there can also be potential gains in terms of enhanced security and privacy. Having evidence-based arguments in favor of a particular approach can help to build trust with your business stakeholders. Having evidence and real data points can help construct an effective business case and address your stakeholder's concerns.

Adoption scenarios are discussed at length as well. We also include a *What not to do* section, to save you some time by covering a trial-and-error type of approach.

Reviewing Chapter 3, Framing Your Adoption

In this chapter, we introduce Microsoft's **Cloud Adoption Framework** (**CAF**), an extensive online resource containing documentation, templates, tools, and ample advice on best practices designed to help any (and every) organization move to the cloud. Starting with business strategy and alignment, we argue that a clear understanding of the business motivations for cloud adoption is critical for overall success. Before embarking on a cloud adoption journey, it is important to identify and align your business goals. You should ask yourself what you hope to achieve by leveraging cloud computing and why it is critical for your organization.

We discuss how to capture your current IT landscape and assess the organization and technology readiness and offer advice on how to start forming an adoption plan. It is essential to take an inventory of all your current IT assets, their usage patterns, and how they can be optimized through cloud services. This will help you determine which of your existing resources need to be retired and which need to be migrated or re-architected into a modern cloud architecture. When transitioning workloads and applications to cloud environments, organizations must ensure that they have adequate security measures in place to protect data and applications from unauthorized access, misuse, or other risks. Thus, understanding regulatory compliance requirements, as well as assessing threats related to data governance, privacy, and identity management becomes important at this stage.

We describe how to approach migration, particularly at the early stages when there are *known unknowns*. One important resource is the CAF readiness assessment. This helps identify any gaps or areas of improvement in an organization's current IT setup, as well as its workforce capabilities. As part of any digital transformation initiative, it is essential for organizations to equip employees with new skillsets so that they can use cloud technologies effectively. By completing a readiness assessment, organizations can gain valuable insight into the issues they need to address before moving ahead with their cloud migration.

By thinking about what a typical cloud adoption journey entails, even after the initial workloads are migrated to the cloud, we hope it helps you uncover key questions that need to be answered for your specific situation. One common topic that can create a lot of questions is vendor management – specifically, how to choose the right partner to help you adopt the cloud. Organizations should evaluate various vendors in order to select the most suitable partner. You may identify knowledge and skills gaps, so bringing in expert consultants makes sense, but also consider managed service providers to take away some of the responsibility of running the cloud. Of course, selection will depend on a number of factors such as business policies, vendor availability, and pricing considerations. Having a strong partnership is key toward ensuring success during any kind of transition process that involves adopting new technologies such as the cloud. Microsoft can help here by introducing and matching potential vendors from their partner network.

Reviewing Chapter 4, Migrating Workloads

Migrating existing workloads to the cloud is no trivial task, but the best advice is to start *doing* as soon as possible and continuously improve as you learn more. There are so many things to consider: which applications to move, how to manage networking, how to deploy applications, and more. Before beginning migration, it is important for organizations to analyze their existing IT infrastructure and understand the relationships between different applications and components. This will enable you to identify potential points of disruption or failure that may occur during the migration process. Regardless, it is sound advice to plan for downtime. With any major IT change, there is a high chance of service disruption. Thus, it is important for organizations to account for this in advance. Do not rush, make sure to test it, and ensure you have fallback options.

There are a few common ways to approach cloud migration. The first is lift-and-shift, which is exactly what it sounds like – moving your existing infrastructure and applications to the cloud without making too many changes (of course, you can never get away from the fact that things need to change). This is the simplest and quickest way to migrate, but it doesn't take advantage of all the benefits the cloud has to offer. Regardless, make sure to adequately test their migrated applications extensively before moving into production/live operations mode. This is especially true if moving from *bare-metal* hardware to a cloud virtual machine. Unexpected errors or bugs could appear in seemingly random situations. It could take some time to find the root cause or underlying conditions to reproduce the bug.

In this chapter, we covered some of the most common cloud migration scenarios, Azure migration tools, and offered tips to make your transition as smooth as possible. We discussed in detail how to migrate existing workloads, as well as the importance of testing your migrations, so that you can safely do them over and over until they are right – or optimized, where applicable. Once an application is fully migrated and running in production mode, organizations should monitor its performance constantly so as to identify any patterns of abnormalities or areas of optimization early on. These metrics could include response times, memory utilization, CPU usage, and so on, which help organizations maintain better control over their cloud-based systems' performance and behavior over time.

However, there's no one-size-fits-all answer when it comes to migrating to the cloud – it all depends on your specific needs and requirements.

Reviewing Chapter 5, Becoming Cloud Native

A cloud-native approach to software development was the main thread of this chapter, together with agility and innovation. *Cloud native* is a very specific way of developing, packaging, and deploying your services, and we discussed the intricacies in this chapter. We also started counting the cloud-native services in Azure. We didn't get to them all, but we gave you enough direction for you to be able to discover more of them as you need them. Being cloud native requires an understanding of cloud technologies such as containers, microservices, serverless computing, infrastructure as code, and the like. Get up to speed on these topics before deciding to become cloud native. But it's worth it – so much so that we even tried to scare you into adopting cloud native by claiming your competitors were doing that as well.

Then, we went on to discuss agility. And if you remember one thing from this book, remember that agility trumps everything else. Only agility gives you a *fail-fast-fail-safe* framework where you are not wildly going on a tangent that no customer of yours cares about and discovering this only when you nearly bankrupted your organization. As part of making your applications cloud native, you should also adopt DevOps practices such as **continuous integration** (**CI**) and **continuous delivery** (**CD**). This will enable faster development cycles, while reducing release times during deployment periods significantly. By regularly monitoring your applications with Azure Application Insights, you can quickly identify issues or areas improvement that can be addressed as part of your CD cycle. This enables failing safely: delivering incremental value more frequently and monitoring usage allows you to pivot quickly when you get it wrong.

Innovation is also something we go into as, in conjunction with agility, innovation must be embraced and adopted as a necessity for the cloud adoption to work, and in more general terms, for your organization to stay relevant in the market. The cloud has revolutionized the way companies innovate and use technology. It is easier for businesses to access powerful computing resources, store data securely, and rapidly develop new applications. The cloud provides a range of advantages that can help organizations stay ahead of the competition. For example, it enables dynamic scalability so that businesses can increase or decrease their usage according to their needs. This means they can quickly respond to changes in market trends or customer demands without needing to invest in new infrastructure. Additionally, cloud services are often cheaper than traditional hardware solutions due to increased efficiency and less manual maintenance. This makes it possible to launch products faster, conduct research more efficiently, and create experiences that delight your customers. But as we said, transforming the business organization into an agile machine is really necessary to capitalize on all the cloud has to offer.

Reviewing Chapter 6, Transforming Your Organization

The only logical conclusion from all the previous chapters is the need for your organization to embrace change. So, transformation is the order of the day – *continuous transformation, every day*.

We discussed how to start and how to guide the transformation, and how to adapt the CAF so you can achieve the right transformation.

We also briefly discussed **site reliability engineering (SRE)** as one of the core principles of running a SaaS platform. We didn't go into nearly as much detail as needed, but SRE practices are a series of books on their own. For organizations leveraging the cloud, SRE can be an incredibly powerful tool for ensuring applications remain up and running, despite any challenges they might face. By adhering to SRE best practices such as automation, monitoring, feature flagging, and continuous integration and deployment, businesses can ensure that their applications are running as they should while reducing the time spent manually managing them. And remember time is money. By leveraging tools such as **Application Performance Monitoring (APM)**, teams can quickly identify issues before they become major problems, thus preventing outages that could have an adverse impact on your customer's experience as well as your business reputation.

But hopefully, the emphasis on SRE in this chapter has been stored in your brain to be triggered at the opportune time – we would not be comfortable running a SaaS platform for an organization without an SRE or red-blue-purple team. You need SRE teams to ensure all that change coming from agility and innovation doesn't disrupt the day to day running of your services for your customers. Yes, you can have both!

Cloud governance was discussed next, as you cannot ensure any reliability without proper governance, you cannot ensure cost optimization without governance, and so on. Governance is an essential part for cloud adoption and will be a key factor in determining whether cloud adoption is successful or not.

We also gave you a definite answer to the question *"Will things ever go wrong?"* by giving you guidance on communicating in an emergency. We gave you this not just in case but rather because this will be a constant need for you going forward – so as always: you'd better be prepared. You cannot have agility and innovation without disruption, and while you can work to be really great at managing disruption, there is always the edge case where disruption wins over governance and management and agility and innovation.

Reviewing Chapter 7, Creating a New Operating Model

Everyone knows that the cloud is where it's at! So if you're looking to up your game, you will need to *go long* on a cloud operating model for your organization. This can mean new roles, new teams, new processes, and certainly plenty of training for new cloud skills. The most important component of any successful cloud operation is the team behind it — so one of the primary steps in designing your cloud operating model should be to identify key stakeholders and roles within your organization who will be involved in setting up and managing your operations on the cloud. Establishing clear workflows and procedures for routine tasks such as cloud infrastructure provisioning, configuration management, or monitoring will help ensure consistent results and make sure everyone on your team is aware of expectations and responsibilities.

Selecting the right tools and technologies based on the specific requirements of your application or service is essential when creating a reliable cloud architecture. Your business needs must play an important role in shaping your cloud – consider factors such as cost savings, operational efficiency, scalability, and flexibility before settling on a target design. With an Azure Landing Zone, you can design a solid foundation for identity management, subscription hierarchies, cloud resource usage policies, and cost management. This will remove the guesswork later and get your organization off the ground with a streamlined cloud management approach sooner.

We ended this chapter with some best practices specific to Azure that should help achieve continued success after the initial cloud adoption journey.

Reviewing Chapter 8, Architecting Your Cloud

This chapter was the culmination of all the advice from previous chapters, but it can be tricky knowing exactly how to architect your cloud for success. You are the only one privy to all the nuances of your organization, your technology stack, and the skills across your organization! Luckily for us, Microsoft has developed the Azure Well-Architected Framework to help ground us in what good architecture looks like. This framework provides five pillars that are crucial for designing scalable high-performing cloud solutions: reliability, cost management and optimization, operational excellence, performance efficiency, and security. Whether you're a newbie in the world of tech or a coding guru, this framework is designed to help you hit all the right notes with your designs and ensure consistent success. Which decisions to make, when, and how to go about this is what we discussed in this chapter. Also, which questions to ask – of yourself and of others in your organization. A key take-away should be that good architecture does not last very long. It is important to establish regular cloud architecture reviews to ensure that applications and workloads remain performant, reliable, secure, and cost effective over time. Utilize tools such as Azure Advisor to regularly assess architecture changes while also documenting results to track progress over time. Use this data to quickly identify areas where performance can be improved, or costs optimized. But also remember to take action! As problems arise or new opportunities present themselves, make sure you act on them swiftly – implement recommended changes and monitor the situation carefully until satisfactory results have been achieved before moving on to new projects!

Summing up

That's all folks!

You would be forgiven to have thought that cloud adoption was an easy task before reading this book. We hope that you now recognize the true challenge that lies in embracing the corresponding organizational and cultural changes. For some, this may mean breaking old habits that have been established for many years and adapting to a new way of doing things. But don't be intimidated, cloud adoption could be your long-awaited chance to use innovative solutions that can deliver business value faster and improve efficiency – and a lot of fun to use! Whatever your reasons might be for cloud adoption, one thing is clear: you need to take into account the full range of challenges involved. Navigating the complexity of this transition requires thoughtful planning and execution to ensure success.

With this book, we have given you a wide scope to consider many challenges and solutions. Fortunately, you have resources such as Microsoft Azure's Cloud Adoption Framework, which provides far more detailed guidance, scenarios, and industry insights – but best of all, is continuously updated as both the platform evolves and also as new learnings and best practices are discovered. Remember, it is important to consider getting expert help where needed to ensure success; you can decide what their level of involvement should be, but it rarely hurts to get an outsider's perspective.

Cloud adoption can be a daunting task, but we hope this book has helped to make it a little less scary. We could not be happier that you decided to read this book, and we would be delighted to hear from you. So, reach out and tell us about your cloud adoption experience and let us know where we got it right, but especially where we got it wrong.

Ciao!

Index

W

www.packtpub.com

Subscribe to our online digital library for full access to over 7,000 books and videos, as well as industry leading tools to help you plan your personal development and advance your career. For more information, please visit our website.

Why subscribe?

- Spend less time learning and more time coding with practical eBooks and Videos from over 4,000 industry professionals

- Improve your learning with Skill Plans built especially for you

- Get a free eBook or video every month

- Fully searchable for easy access to vital information

- Copy and paste, print, and bookmark content

Did you know that Packt offers eBook versions of every book published, with PDF and ePub files available? You can upgrade to the eBook version at packtpub.com and as a print book customer, you are entitled to a discount on the eBook copy. Get in touch with us at customercare@packtpub.com for more details.

At www.packtpub.com, you can also read a collection of free technical articles, sign up for a range of free newsletters, and receive exclusive discounts and offers on Packt books and eBooks.

Other Books You May Enjoy

If you enjoyed this book, you may be interested in these other books by Packt:

Azure for Developers

Kamil Mryzglód

ISBN: 978-1-80324-009-1

- Identify the Azure services that can help you get the results you need
- Implement PaaS components – Azure App Service, Azure SQL, Traffic Manager, CDN, Notification Hubs, and Azure Cognitive Search
- Work with serverless components
- Integrate applications with storage
- Put together messaging components (Event Hubs, Service Bus, and Azure Queue Storage)
- Use Application Insights to create complete monitoring solutions
- Secure solutions using Azure RBAC and manage identities
- Develop fast and scalable cloud applications

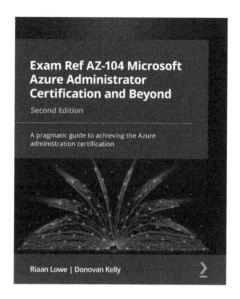

Exam Ref AZ-104 Microsoft Azure Administrator Certification and Beyond - Second Edition

Riaan Lowe, Donovan Kelly

ISBN: 978-1-80181-954-1

- Manage Azure Active Directory users and groups along with role-based access control (RBAC)
- Discover how to handle subscriptions and implement governance
- Implement and manage storage solutions
- Modify and deploy Azure Resource Manager templates
- Create and configure containers and Microsoft Azure app services
- Implement, manage, and secure virtual networks
- Find out how to monitor resources via Azure Monitor
- Implement backup and recovery solutions

Packt is searching for authors like you

If you're interested in becoming an author for Packt, please visit `authors.packtpub.com` and apply today. We have worked with thousands of developers and tech professionals, just like you, to help them share their insight with the global tech community. You can make a general application, apply for a specific hot topic that we are recruiting an author for, or submit your own idea.

Share Your Thoughts

Now you've finished *Azure Cloud Adoption Framework Handbook*, we'd love to hear your thoughts! Scan the QR code below to go straight to the Amazon review page for this book and share your feedback or leave a review on the site that you purchased it from.

`https://packt.link/r/1803244526`

Your review is important to us and the tech community and will help us make sure we're delivering excellent quality content.

Download a free PDF copy of this book

Thanks for purchasing this book!

Do you like to read on the go but are unable to carry your print books everywhere? Is your eBook purchase not compatible with the device of your choice?

Don't worry, now with every Packt book you get a DRM-free PDF version of that book at no cost.

Read anywhere, any place, on any device. Search, copy, and paste code from your favorite technical books directly into your application.

The perks don't stop there, you can get exclusive access to discounts, newsletters, and great free content in your inbox daily

Follow these simple steps to get the benefits:

1. Scan the QR code or visit the link below

https://packt.link/free-ebook/9781803244525

2. Submit your proof of purchase

3. That's it! We'll send your free PDF and other benefits to your email directly